The Atom and the Apple

The Atom and the Apple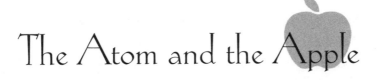

Twelve Tales from Contemporary Physics

Sébastien Balibar

Translated by Nathanael Stein

PRINCETON UNIVERSITY PRESS • PRINCETON AND OXFORD

Black Night

I was nine years old. My father was teaching math and my mother classics, each in different high schools in Tours. Just after the war, they found a pair of jobs in a quiet city, which they actually didn't like very much. So, naturally, they decided to escape it as often as they could, and to look for a house where they could take their four children on vacation. They had no money, but they did discover an old farm in ruins at the foot of a very pretty village in Provence, with fourteen rooms surrounding a courtyard with a well, on five hectares of pebbly terrain where peasants had once tried to grow wheat. At the time, the region didn't have running water, the yield from the land was minimal, and tourism was non-existent. But, to my parents' eyes, the place was perfect. They bought this marvel for a mere 450,000 old francs—barely a few months' worth of their teachers' salaries.

The first winter we spent in the few inhabitable rooms was glacial, but hopeful. And by the following fall, I had decided on my future career: I wanted to work for the department of Eaux et Forêts.[1]

I should mention that the government bureau of Eaux et Forêts was encouraging reforestation. With their help, we planted several thousand cedars, several varieties of cypress, Aleppo and Austrian Black pines, some lindens, acacias and poplars, a hackberry and a catalpa, a rowan, one hundred forty almond trees, thirteen rows of vines including Angevin and Cardinal varieties, a Chasselas, a Muscat, a Gros Vert and a Dattier . . . And I do mean "we," since despite being only ten years old by that time, I had taken up the care of our plantings with my father. Among the pebbles and thistles, I stubbornly tended the attempts of our trees to grow, though drought in both winter and summer decimated a number of them each year. However, with steadfast encouragement, about half of them survived, managed to reach the water table within about ten years, and then suddenly began to grow ten times as fast. The vegetation on the ground was transformed; the thistles disappeared, and the pebbles were quickly covered with moss and a sprinkling of pine needles—favorable conditions for porcini mushrooms in autumn. One day we even found some morels; and we met rabbits, hoopoes (a type of bird), and various kinds of butterflies, which I collected. I made picturesque paths through the new forest we had created and proudly decided that nature was more pleasant when managed than wild.

But then water came to the whole region, thanks to the Durance-Ventoux land management plan: the Renaissance castle in the village started exhibiting ugly geometrical paintings, Madame Francine Coeurdacier sold her food shop to Hediard (an expensive grocery chain), tourists flowed into the region and covered it with swimming pools, and the formerly dark alleys were lit with yellow light . . . Fortunately, our forest acted as a screen to all this excitement, and our nights stayed dark, pure, and perfumed; only the crickets and the toads broke the silence.

Fifteen years later, our parents left the pleasures of this paradise to us. It was then that our friend Jean-Pierre Maury brought us a telescope. Jean-Pierre liked woodlands, astronomy, big Cuban cigars, and popularizing science. From the garbage bins populating the Jussieu campus at the University of Paris, where he taught physics, he gradually extracted a scrap-metal frame, a wide PVC tube 25 centimeters in diameter, an old electric motor, a large aluminum disc, and three wheels. When, to all this bric-a-brac, he added a few planks of light oak and an eyepiece from a microscope, as well as a large parabolic mirror which he had polished himself in his kitchen and had had silver-plated in a workshop in the Marais, Jean-Pierre had successfully put together a very respectable instrument for observing the sky; and he gave it to us as a present.

Thus equipped, we began to pretend that we were all miniature Galileos.

We took the telescope out onto the former wheat field of our farm, well-shaded from neighboring lights by our pines and cedars, turned the axis of rotation to make it parallel to the Earth's, and began endless observations.

My first observations of the night sky were simple: when it was full, the dazzling Moon filled the entire field of vision of our homemade telescope. But I preferred looking at the first quarters of the Moon, when the shadows of the craters would spread out beyond the boundaries of the lit zone, and when we were able to distinguish the neighboring penumbra.

Then I figured out how to find Jupiter. You only had to look south and imagine the place where the sun had passed during the day—a sort of great circular trajectory, which begins in the east, rises to the south, and descends again to the west. The planets follow the same path because all of them, including us and the Sun, are on the same plane. The trajectory of the Sun and the planets in the sky is the intersection of the plane of the ecliptic and the celestial vault. If I found a very bright object in this region of the sky, there was a good chance that it was Jupiter, unless it was Venus instead.

Jupiter was easy to recognize since, like Galileo in 1610, I could clearly see four satellites aligned with it. I learned their names: Io, Europa, Ganymede, and Callisto; then I perceived that these four shining points changed places from one day (or rather, night) to the next. Well, of course: they travel around Jupiter, and I would sometimes see two on one side and two on the other, then three on one and only one on the other, and so on. If the sky was really clear, I could even distinguish the dark bands formed by Jupiter's atmosphere, torn apart by abominable winds. The bands are parallel to the equator and in line with the satellites: the whole thing turns along a single plane! As for Venus, which I found near the Sun—that is,

3

towards the western sky in the early evening or towards the east in the early morning—it was also easy to recognize: a very bright object, but with no satellites. And I discovered that Venus sometimes even had a crescent shape—like the Moon, only smaller.

I was able to tease my friends: that big "star" wasn't a star at all, it was a planet! We could easily tell that it didn't emit its own light, but rather was only reflecting the light it received from the Sun, and that we were obviously looking at a nearby object, since we could clearly see its diameter. I felt small, overcome with vertigo in the midst of this celestial merry-go-round. As I kept adjusting the direction of our telescope, the motor of which never seemed to work very well, I continued to think about Galileo, convicted for having defended science against the Catholic Church. I didn't know at the time that he had actually never added under his breath, "and yet it moves,"[2] as the legend goes—that he didn't have enough strength to do it by the end of his trial; but I did understand that what I was perceiving in my telescope eyepiece was the rotation of the Earth on its own axis, not its revolution around the Sun. That revolution is what makes Jupiter change locations in the sky all throughout the year, unlike the stars. I began to think about the change of seasons, about the inclination of the rotational axis of the Earth in relation to the plane of the ecliptic, about the Ice Ages and the interglacial periods that the Earth has known in its long history . . .

In my daydreams about relative motion, Galileo was my constant companion. In effect, what difference is there between "The Earth revolves around the Sun" and "The Sun revolves around the Earth"? If there were only two objects, the Earth and the Sun, these two scenarios would be strictly equivalent. For example, is it the Moon that circles around the Earth, or the Earth that circles around the Moon? As proof of our general egocentrism, we tend, of course, to say that it's the Moon that revolves around the Earth: "We, the Earth, exert the greater force, which makes the little satellite revolve." But in fact each one revolves around the other, and the Moon isn't really that small compared to the Earth anyway.

In order for it to be the case that no one today says that the Sun revolves around the Earth, I realized then that it had taken generations of astronomers observing not only the relative motions of the Earth and the Sun, but also those of the stars and the other planets. Now, the stars are always in the same position in the sky from one year to the next, which isn't true of the planets—even from one week to the next we see the planets move in relation to the stars. When Copernicus judged that all the planets, including the Earth, moved around the Sun, we finally had a simple model at hand which allowed us to predict the positions of all the objects in the sky. The roles of the Earth and the Sun were no longer symmetrical. In 1915, after a few anomalies in the movements of Uranus and Neptune were observed, it was even predicted that there must be an additional planet influencing those two. In 1930, Pluto was discovered. So it would seem ridiculous today to claim that the Sun revolves around the Earth, which would then be the fixed, immovable center of the Universe. At the end of this series of personal revelations, I felt smaller and smaller, not only within the physical Universe but also in the universe of scientific ideas. I had gained a whole new respect for the history of astronomy and its great men.

Without getting carried away, I went back to conquering the heavens. I perceived another shining point, a little reddish, on what ought to have been the path of the planets. Mars? A glance was enough for me to see that this reddish object also had a visible diameter—that it was indeed another planet. Mars gets its red color from the strong concentration of iron oxides in its grainy soil, and scientists are wondering more and more seriously if it once had life, since subterranean and icy water has been discovered there.

Combining my love of nature with a strong dose of rationalism inherited from my father, I became a physicist. But some days, I'm tempted to buy myself a really nice telescope and go back to Provence, far from the lights of the city and its murky atmosphere, so that I can see for myself that Mars has a white cap which travels from its north to south pole according to the Martian seasons. This ice cap exists, because there is snow made of CO_2 there, and carbonic snow is white (like the H_2O snow here on Earth, only a bit colder). And since I know where to find the ephemerides in my newspaper—or better yet, on the Internet—I can know whether

Saturn is visible at a decent hour and in what part of the sky. I remember its iconic rings, which I would try to count as a child, and the inclination of which varied from one year to the next. I also seem to remember that one day, Saturn, sitting inside its rings and seen from one side, looked like an eye—an eye that was looking back at me.

Really, there's only one step from science to dreams, or vice versa.

Recently, I came across this question, which at first seemed disarmingly naïve:

"Why is it dark at night?"

How could someone ask such a question, you might think? The truth of the matter is that there are few stupid questions, and the most naïve are often actually the most fundamental in nature. That is definitely the case with this question, since it's related to the origins of the Universe, to the theory of the Big Bang, and the current frontiers of cosmology.

Of course, at night, the Sun is shining on the other side of the Earth. But since all of the other stars are just so many other Suns, so why don't they shine enough for us to see at night just as we do during the day?

Obviously, they are far away, and we know very well that the farther we are from a source of light, the less we're lit. We get four times less light from a beacon two kilometers away than from one which is only one kilometer away.

Still, there are a great many stars. If the stars don't shine brightly enough upon us for us to be able to see clearly at night, perhaps it's because there aren't enough stars in the sky? In 1823, this excellent suggestion led Wilhelm Olbers, a German astronomer, to a remarkable conclusion: the observable Universe is not infinite. Now, the problem of the finitude of the universe had been bothering philosophers for millennia, though today, we know the answer to this riddle. Was Olbers right? To find out, we need to go back a few steps.

How many stars are there in the sky? It's not so easy to answer this prefatory question. For simplicity's sake, imagine that the average density of stars in interstellar space is constant.[3] Now imagine counting the number of stars shining at a certain distance from the Earth. Then imagine doing the calculation again for double the distance: the number of stars will be four times as large, but, since they shine from twice as far away, we receive four times less light from each of them. To sum up, the light that comes from twice as far is precisely equal to that coming from twice as close, since the increase in the number of stars is compensated by the decrease in light received as a function of the distance. Next, Olbers calculated the total light we receive from all of the stars in the sky, from the nearby stars out to the faraway ones. If the faraway ones were infinitely far, we should get an infinite quantity of light; but, this isn't the case. Thus we have to conclude that we do not see stars out to infinity— we do not receive light coming from more than a certain distance. This limit must, certainly, be far away, but it can't be infinitely far. Olbers showed, then, that the Universe has a horizon. But how?

Let's keep thinking. As you know, light propagates at 300,000 km/s, which is a lot, but not infinite either. To reach us today from a source infinitely far away, the light would therefore have to have been emitted an infinitely long time ago. Therefore, if we don't receive an infinite quantity of light, perhaps it's because the visible Universe isn't infinitely old! Now that is both more precise and more correct. It is dark at night, because the stars have birthdays; they haven't always been there.

Reaching such a conclusion was truly a revolution. But our conception of the Universe has been greatly enriched since then. In 1923, in effect, a century after Olbers, the American Edwin Hubble discovered that the Andromeda Nebula is actually a galaxy beyond our own.[4] He then discovered other galaxies, succeeded in measuring their distances from us, and then—and this is his greatest discovery—he saw that all of these galaxies were getting farther away from us at a speed proportional to their distance. So he confirmed, in 1929, a hypothesis that was formalized in 1927 by Georges Lemaitre, and that would eventually become famous: the Universe must be expanding as a result of a powerful initial explosion.

Such an impressive phenomenon deserved a nickname, and so it was called the "Big Bang." But what made Hubble come up with the idea that the Universe had been flying off in all directions since the initial explosion? Progress in this area of study resulted from the fact that we were able to measure the speed of the stars.

When I was little—clearly my childhood had a big impact on me—I had miniature cars with which I held imaginary races. My early passion for fast cars has happily abandoned me, but at the time, I wasn't content just to push them along roads drawn with chalk: I made the sounds, as well. All that was missing were the lights. Passing in front of me at high speed, the cars went *eeeeeeeeeeeeaaaaaaaaaooooooooo*. Like all my friends, I knew from experience that real cars getting closer to me had a higher sound than when going away, but I didn't know that this effect had been predicted by the Austrian physicist Johann Doppler in 1843, and was then tested in a rather amusing way, by placing trumpet-players on a train. I also didn't know that Doppler had predicted that light, being a wave like sound, would show the same effect, nor that the French astronomer Armand Fizeau had deduced from that a way of measuring the speeds of stars in relation to us.

In stars, each sufficiently hot atom emits light which has a particular structure: it is made up of a collection of rays of different colors (called a spectrum), which is a veritable signature of each element in the star. We therefore know how to chart the chemical composition of each star by analyzing the signatures of these atoms in the stellar spectra. Now, these spectra, these colorful signatures, are all more or less shifted towards the lower frequencies, that is, towards red, because of the so-called Doppler-Fizeau effect. So, all the stars are getting farther away from us. By measuring this shift (which is called, appropriately, the "red-shift"), we find how quickly each star is moving away from us. Some are very far and moving away very quickly, and they get very red because of it; others are closer and barely move relative to us, and don't blush so much. So, imagine we were going back in time: all the stars, and all the galaxies of stars, converge at a point which is the origin of the Universe. Hubble thus imagined that the current Universe resulted from an initial explosion and hadn't stopped expanding since. To date, this is still the central hypothesis of cosmology. The theoretical model has been further developed, and the measurements in support of it have become more precise: we think today that this Big Bang took place 13.7 billion years ago. This is the age of the Universe, three times the age of the Earth and the Solar System, which has also been measured, but by a different method. I'll come back to this; but did you know that the Universe is three times older than the Earth?

Since one unlikely question has taken us as far as the Big Bang, it would be a shame to stop there. So, I'll ask another:

"Is the sky going to fall on our heads?"

I don't know if the history books for primary schools still associate this question with the supposed naïveté of the Gauls, as they once did in France. It is, however, a very real and serious question for modern cosmology, and anything but naïve.

The Sun attracts the Earth, which, as everyone has known since Newton, attracts the falling apple: that's universal gravitation. The stars, therefore, also attract each other. The force of attraction depends on the distance between and the mass of each object, and is the prime mover of the dynamics of the Universe. Imagine, then, a large concentration of masses somewhere, a sort of dense cloud: it will eventually collapse upon itself, and over the course of this implosion, heat up so much that nuclear reactions will be set off. That's how stars, which are nuclear furnaces, are born and shine. One or two are born per year in our galaxy alone, which isn't a comparatively young one.

But what about the Universe itself? If it's very dense, it should collapse upon itself in the same manner, given how much each star attracts every other star. And if this attraction is strong, couldn't the expansion resulting from the initial explosion reverse itself one day? Is it possible that after an expansion phase, the Universe will retract, the velocities of the stars will reverse themselves, and everything will collapse back in upon itself? If that were true, some day someone should witness (someone, not us—it will be a long time coming) another singular event, a sort of Big Bang in reverse which the physicists have given another nickname: the "Big Crunch."

When you throw an apple up into the air, it falls back down. Will the Universe fall back upon itself? Is the sky going to fall on our heads?

Not to worry. Astrophysicists have made a lot of progress in this area and have just discovered that it's the opposite: the Universe is flying away at a faster and faster rate; its expansion is accelerating, and the sky will surely not fall back down on our heads. But how did we manage to establish that fact? It's a long story.

The first thing needed to figure all this out is some idea of the density of matter in the Universe. Visible matter is very diluted: the equivalent of one hydrogen atom per cubic meter. That's not enough to reverse the current expansion. But, by studying the rotation of the galaxies, and then the movements of the clusters of galaxies, we perceived that there must be a lot of matter that we don't see. What could this invisible matter be made of?

First of all, let's consider the planets that revolve around stars. This is matter which doesn't inherently emit visible light; however, we can detect the presence of these dim planets around their stars indirectly. For example, we can observe partial eclipses of stars: their light decreases slightly when one of their planets passes between them and us. We can also observe that the movement of certain stars oscillates as a function of the rotation of their planets; in fact, that's how some planets of our own solar system were discovered. There is also a lot of gas and dust between the stars. However, all this ordinary matter, shining or not, only represents about 20% of the total matter in the Universe.

The candidates for the missing 80% include "neutrinos," elementary particles with which we know the Universe is filled, and the mass of which we have just discovered is not zero. There may also be other unknown particles which also have mass. All this supplementary matter, which is not made of protons, neutrons, and electrons like ordinary matter, is what we call "dark matter," not only because it doesn't shine, but also because we don't really know what it's made of. Why are we convinced that it exists? Because, like planets, we've been able to observe it indirectly. Without it, the angular velocity of some galaxies would be incomprehensible—unless the laws of gravitation were completely revised . . . But I won't go there. What intrigues all the physicists in the world is that this poorly understood dark matter represents the vast majority of the matter present in the Universe. Now, that opens up some big avenues of research—all the more so, since the mysteries to be penetrated run much deeper than even the fundamental nature of dark matter.

The possibility of a Big Crunch depends, in effect, on other surprising aspects of the Universe. In particular, it's possible that the Universe is curved. In his Theory of General Relativity, Einstein represents gravitational attraction as a curvature in space. To understand what that means, physicists are in the habit of comparing space to a mattress, even though one might find the allusion to be a bit simplistic. Lying on a soft mattress, a body distorts the mattress's shape, creating a local curvature—a hollow, if you like—and any neighboring body is attracted towards this hollow. In this analogy, space is two-dimensional; it is a distorted plane. But the space of Einstein's physics is actually four-dimensional: three dimensions for ordinary space, plus the additional dimension of time. Picturing the curvature of four-dimensional space isn't easy, hence the usefulness of the mattress. In any case, we represent gravitational attraction as the result of local deformation of space by the masses located there (stars, galaxies, and so on).

Allow me to make a parenthetical remark. Maybe you believe that the curvature of the Universe by its masses is just an abstract fantasy? A pure speculation on the part of Einstein which has no relation to perceivable reality, or at least no practical utility? Not a bit! By measuring the duration of eclipses of Mercury, and by observing gravitational mirages in the sky (the relativistic equivalent of optical mirages which you may have observed on a hot road reflecting the light of the sun), it has been verified that the deformation of space makes light deviate when it passes near a star. General relativity also predicts that the time shown on a clock depends on its altitude, since even the Earth curves the space in its neighborhood, and that space has a temporal dimension. Of course, this effect is very weak, but for certain applications which require extremely precise measurements of time, like satellite locating systems

11

(the GPS used by navigators or hikers), it's necessary to take into account the fact that clocks on Earth run more slowly than those on board a satellite. So modern physics is bizarre, but also quite useful. Let me also quickly add a few words about black holes, which are often on the covers of scientific journals and which have also become an observed reality. If the concentration of matter is too high, the structure of space gets more than deformed: it collapses, and it forms a sort of hole from which even light can't escape. At the center of our galaxy, such a black hole exists; its mass is about 2.6 billion times that of the Sun and it swallows stars. We now have photographs of this black hole, taken every six months, where we see stars circle around and get swallowed like boats in an immense mythical vortex, like Edgar Allan Poe's maelstrom.

The curvature of space by matter is therefore a reality. One might then wonder whether the entirety of space itself in fact has a global curvature. Placed on a bulging surface, marbles will all go away from each other, whereas if the surface has a trough, they will assemble in the middle. Any curvature of the Universe could thus influence its future, the acceleration of the expansion or its possible deceleration before a retraction phase. In 2001, this curvature was measured by studying what is called the "cosmic microwave background" radiation of the Universe. We'll see what that means later, but I'll give the result right away: the curvature seems to be nil—the Universe is not curved. I don't know whether you will find that information reassuring. But in any case, it's a little less difficult to imagine than the contrary.

So what is this "background radiation?" Right after the Big Bang, the Universe was so concentrated that it was extremely hot, which separated the electrons from the ions, with the consequence that the Universe was a conductor of electricity, like a metal. That made it opaque, again like a metal. Light was only able to propagate freely in the Universe when the temperature dropped enough for the positive ions to capture the electrons and form neutral atoms. That took place 380,000 years after the Big Bang, the day light began in the Universe. In 1946, George Gamow had predicted that as the Universe expanded, the radiation must have been diluted and cooled, but that there must remain a trace of it—background radiation—a sort of cold light that one should be able to observe between the stars. And in 1965, two American astronomers, Arno Penzias and Robert Wilson, discovered it! This was obviously stunning proof of the validity of the Big Bang model, but I won't go on at length about the predictive power of physics. Today, thanks to detectors that are cooled with liquid helium and loaded onto satellites, we've measured the characteristics of this radiation: everything behaves as it would if the temperature of the background of the sky was 2.73 degrees Kelvin (about −270 degrees Celsius). This background radiation resembles the light emitted by a piece of red iron, except that at 1,000 degrees, iron emits light which is very red, whose wavelength is around 0.6 microns; the background of the sky, on the other hand, is very cold and emits very weak microwave radiation (in between infrared and radio waves), about 1 mm in wavelength. The measurement of the curvature of the Universe comes from a close study of this background radiation or, more precisely, its slight inhomogeneities. We know today that this curvature is practically nil. That wasn't at all obvious, however, especially since when linked up with other measurements, this leads one to a new, and no less surprising claim.

Physicists have, in effect, discovered the existence of another force influencing the expansion of the Universe and which comes from what has been called "dark energy," not to be confused with "dark matter." Maybe you have the impression that, despite being a professional physicist, I'm confusing forces and energy. Rest assured, I know that a force is derived from energy, that they are therefore two quantities that are related to one another, though not identical. Regardless of terminology, what is this dark energy? This force, or rather stress, to be more precise, tends to push the stars away from one another, to expand the Universe, like a sort of negative pressure. I will come back to this notion of negative pressure by discussing the height of trees (chapter 8), since I know that this concept disturbs even professional physicists, and as it happens, part of my own research is concerned with it. Let's just think for the moment that if we compress a gas in a cylinder with a piston, we are applying a positive pressure to it, which of course tends to push the atoms or molecules inside closer together. If we imagine a reverse action, pulling on the piston instead of pushing, we are expanding the gas and spreading out the molecules. The problem is that we can't apply negative pressure to an ordinary, classical gas: it's impossible. We can reduce the pressure to less than regular atmospheric pressure (+1 bar), but as the expansion progresses, the gas gets more and more diluted and its pressure and density go to zero. That is the ideal gas law, a consequence of the absence of interactions between the molecules and atoms which constitute the gas. On the other hand, applying negative pressure to a solid or a classical liquid is easy: it means pulling rather hard on it. We will see that the pressure is negative in the sap high up in tall trees. But the Universe is more like a sort of gas composed of stars, rather than a solid or liquid, in which a strong cohesion among the molecules is the rule.

In order for the Universe to be under negative pressure, we need there to be a non-classical cause, a quantum one, in fact. It gets tricky to explain, but that's not astonishing: this is what a number of researchers are now coming up against. We've known since the beginning of the twenty-first century that the existence of dark energy implies a force pulling on the Universe; we know that the effect of this dark energy on the dynamics of the Universe is three times as intense as that of all the matter in it, visible or not; and we know, finally, that as a consequence of dark energy, the expansion of the Universe is accelerating, and has been doing so for 5 billion years, whereas without it the expansion would necessarily slow down. Measurements consistent with this have been done, not only on the background radiation, but also on the explosions of certain stars and on gravitational mirages. So we are certain that this dark energy exists, but we don't know yet where it could have come from. Some invoke what they call "quantum fluctuations of the vacuum," but that would yield a colossal amount of dark energy, bearing no resemblance to observed reality. So it's something else. Is there an unknown symmetry in the Universe? Is it what so-called "supersymmetric" theories, which attempt to unify general relativity and quantum mechanics, are starting to predict?

Here we've arrived at the vanguard of contemporary research, a frontier of knowledge where physicists admit to not knowing, not understanding. We have learned so many things about the Universe in one century that the questions asked today are even more mysterious than the ones being asked before.

One more word about Einstein to end this chapter. It just so happens that it was he who first had the idea of such a mysterious force. He introduced it in 1916 in the form of a constant, called the "cosmological constant," which he added to the equations describing the dynamics of the Universe. This fact is interesting because at the time, the theory of the Big Bang didn't exist yet, and Einstein thought the Universe was stable. However, since the stars attract one another, his model needed a force to prevent the Universe from collapsing on itself. Einstein therefore introduced his constant to represent what we now call "dark energy." Later, he felt that this constant was contrived, that it had no physical justification, and that his introduction of it had been the mistake (the "biggest blunder") of his life. Today, thanks to Einstein (and perhaps in spite of him), the cosmological constant has resurfaced.

When I think about it, I have the impression that we all tend to believe that the Universe, nature, life on Earth—all of it—is stable and benevolent, and unless we destroy this beautiful balance which has been granted us, the future won't be catastrophic. Behind these ideas, there must be the belief that the world must have been created to last. And yet, it must be admitted that this view is false: the Universe isn't stable, and the Sun and life on Earth will necessarily have an end. That's no reason to hasten the end, but life, and especially human life, is only a passing episode in the long history of the Universe.

My Cousin the Leek

January 21st, 2001.

I was having lunch yesterday with my friend Pascale, who is a neuropsychiatrist. She was talking to me about the Rorschach test. You put some ink on a sheet. You fold it in half, then unfold it. And you end up with a symmetrical patch. Then you ask people what the patch makes them think of.

In her hospital department, they keep the results and analyses from old tests, as well as the details of the people who were tested, including a way of finding them again. Later, they hope to find out what became of these people, and compare them with the psychoanalytical analyses and predictions that had been made years before.

"Well done!" I thought. Here was a beautiful example of scientific methodology at work, and even more impressively, these psychiatrists had the courage to meet face-to-face with evidence of any potential mistakes they might have made during their first analyses!

We shared a smile over the various scenarios that could unfold as a result of these experiments. Besides the human interest side of these experiments, it is likely that they will serve the purpose of helping to bring possible weaknesses of these Rorschach tests to light, and help the profession make progress.

I remember when I had to take a Rorschach test during my mili-

Figure 1. One of the symmetrical inkblots that Hermann Rorschach created while preparing his psychological test.

tary service; I really wonder now what diagnosis the psychologist colonel who was directing the program could have made. No one ever discussed it with me, and so I wondered aloud what the point of giving me the test was, precisely. Was it anonymous, geared towards research on military psychology? Or were the names kept on file in the archives of some kind of Big Brother?

We exchanged another smile.

Since we were firmly in each other's confidence, I took the liberty of asking Pascale a seemingly critical question:

"How is it that we can really test people's psychologies with symmetrical blots?"

Symmetry is far from insignificant for all of us. Our bodies are just about symmetrical, in appearance, at least. In a mirror, we see clearly that our left halves resemble our right halves.

Resemble . . . But that's just it: we have right hands and left hands which aren't exactly alike. We know quite well that we can't put a left-handed glove on a right hand. So our bodies are made up of two halves that are symmetrical (with respect to a plane) without being identical, a left half and a right half.

17

We can admire the harmony of an ideal, perfectly symmetrical body, or on the contrary, take pleasure in knowing that the human body isn't exactly symmetrical. Other psychological researchers have made synthesized portraits on a computer by completing one true half of a face with another half calculated to be exactly symmetrical. The too-perfect result seems a bit strange. The eye, with the brain in the background analyzing the images, is a curious and impressively precise machine, since it can detect very minor deviations from symmetry, but it is also subject to the power of the unconscious. I imagine psychologists think about this kind of question all the time . . . but Pascale didn't tell me how they try to resolve these kinds of issues.

Then I asked her another question:

"Did you know there's a big puzzle regarding the symmetry in living things?"

As she didn't seem to see what I was getting at, I explained how, apparently, we are all cousins. Not only are we humans all of a family, but we're also related to leeks, sharks, and porcini mushrooms, or girolles, if you prefer.

Among the molecules found in living beings, many are not symmetrical. By that, I mean that they are not identical to their mirror image. They are "chiral," that is, they are like my left hand, the mirror image of which is a right hand.

There are, therefore, left-handed and right-handed molecules. They include DNA, the very long molecule which twists around itself to form our chromosomes. DNA is the basis of heredity, but if I start talking about that, I'll lose track of this business about left and right. We'll have the chance to come back to that later.

What interests me here is that DNA is like a string with two strands, wrapped around each other in a double helix. A corkscrew is also a helix, and it has a direction. To insert it into a cork, you usually need to rotate it to the right, unless it's a corkscrew for lefties. Just as there are corkscrews for righties and others for lefties, there are thus right-handed and left-handed helices. In any case, DNA is a right-handed helix! That's true not only for our human DNA, but also for that of all living beings. There are, in addition, a number of chiral molecules, such as amino acids. But you never find both forms: for each amino acid or other biomolecule, either living beings all have either the left-handed kind or they all have the right-handed kind—nature made the same choice for everyone.

This asymmetry in life was discovered in 1844 by a German chemist, Misterlich, then carefully studied by Pasteur in the crystals that make up the crusty deposits, or tartar, that one finds in wine casks. These crystals have very special optical properties: they rotate the polarization of light to the right. But in 1848, when Pasteur synthesized crystals of tartar in a test tube, he found both left-handed and right-handed crystals, the shapes of which were mirror images of each other. Finding as many right-handed as left-handed crystals in a chemical synthesis seemed just right, since the atoms have exactly the same chemical bonds in the molecules, whether the latter are right- or left-handed.

But if that's the case, why is it that these biomolecules are, to put it simply, right-handed in vivo (in living beings) but right- and left-handed in vitro (in a test tube)? The answer is generally believed to be related to the origins of life on Earth.

Life appeared on Earth almost 4 billion years ago, under conditions that are not well understood, but it is widely believed that we are all descended from one initial single-celled organism. If many organisms had all appeared at the same time, in different places, there would be no reason for there not to be some right-handed and some left-handed living beings—in terms of molecules, I mean. Nor does there seem to be a reason why left-handed beings would have had better chances of survival than right-handed beings. So most scientists believe we are descended from the same common ancestor. Several billion years ago, this single-celled ancestor must have begun to reproduce itself, that is, to live. I leave to the biologists the task of explaining the mutations it must have undergone afterwards in order to yield the different branches of the great tree of evolution, with its plants, its mushrooms, its mammals, its fish, its varied insects. These biologists have even given it a name: LUCA, for "Last Universal Common Ancestor."

That's a far cry from Adam, and of course Eve, supposedly born from Adam's rib (unless we think of Adam metaphorically, of course). The reality which scientists have uncovered, and which they are also working hard to verify, is much simpler and more fascinating: we are all cousins with leeks, sharks, and porcinis, since we have the same distant ancestor. I wonder how many people know or think about this. I fear there are many more who believe too literally in the myth of Adam and Eve, or who don't wonder about their true origins at all.

The equivalence between right and left was thus interrupted by the appearance of life. Life broke the right-left symmetry.

What does that have to do with physics? Just as we wonder why living things are made up of right- and left-handed molecules (I'm simplifying again), we also wonder why the Universe is made of matter and not another thing that is symmetrical to matter, called antimatter. Maybe you think that antimatter is just a bit of fantasy from science-fiction novels? Or that everything possessing mass is matter and vice versa? We had to revise these beliefs in the twentieth century, and the story of that revision reveals the astonishing ways in which physics sometimes makes progress, by surprises and leaps of the imagination.

One of the first surprises happened around 1928 when Paul Dirac, an English physicist born in Bristol in 1902, and future holder of the same chair as Newton at Cambridge, wrote the equation that now bears his name. Relativity and quantum mechanics were still in their youth. Einstein had overturned our ordinary conceptions of space and time, established a relation between mass and energy (the famous equation $E = mc^2$), and affirmed that the laws of classical mechanics, those of Galileo and Newton, needed to be modified once the speed of the objects being considered approaches a universal constant, namely the speed of light. (I hope I will be excused for summing up such a major scientific revolution in so few words.) For its part, quantum mechanics had undertaken to describe the behavior of particles as though they were waves, which was no small thing, but its central equation, Schrödinger's equation, was only valid if the speed of the particles was small compared to the speed of light. But the electron in a hydrogen atom moves fast, and therefore must be described not only by quantum mechanics but by relativity as well. So Dirac wrote a generalization of Schrödinger's equation in order to be able to describe the behavior of rapidly moving electrons.

Among the solutions of his equation, Dirac found two that corresponded to what he was looking for at the outset. These were electrons whose magnetization—what is called their "spin" in quantum physics—was pointing either up or down.

Then, Dirac found something else. His equation contained more than he had put into it at the outset. Another type of particle appeared in his solutions, which Dirac had trouble understanding. Either these new particles had negative energy, which was difficult to conceive of, or their charge was the opposite of that of electrons—that is, positive. At first Dirac believed that these positively charged particles were protons, but after some remarks on his work were made by Robert Oppenheimer and Hermann Weyl, he argued in 1931 that these other particles had the same mass as electrons (2,000 times less than protons), but had a positive charge.

Mathematical fantasy? Theoretical wanderings? Not at all! Only one year later, in 1932, Carl D. Anderson discovered these new, antielectron particles experimentally, which he baptized "positrons." They are produced in the upper atmosphere by cosmic rays, that is, radiation coming from the Sun. Thus, it was true that not everything with mass was necessarily matter; it could be antimatter, a sort of matter with an opposite charge.

Richard Feynman later described these particles in a more general formalism which represented anti-particles as particles—except that they move backward, rather than forward, in time.[5]

So, after forcing us to consider antimatter as symmetrical to matter, except for electrical charge, physicists then made us consider the mirror image of ordinary time: time which flows backwards. Feynman also explained how, when matter and antimatter meet, they annihilate each other and produce a very high-energy kind of light: "gamma ray photons." Conversely, high energy light could give rise to mass in the form of one particle of matter and one particle of antimatter. The distinction between light and matter was disappearing.

21

This astonishing phenomenon, which proves yet again that matter isn't what we thought it was, since it can appear and disappear, transform itself into light and back again, has been the subject of numerous experimental verifications. In order for it to take place, events involving high amounts of energy need to be studied: this was achieved in 1955 in a particle accelerator at Stanford University. Today, all of the anti-particles which are symmetrical to the known particles have been discovered. Some particles, like photons—the quanta of light—are their own anti-particles, but there are anti-electrons, anti-protons, et cetera. Recently, an antiatom of hydrogen was successfully produced, with a positron revolving around an antiproton, and we would like to know whether it falls at the same rate as an ordinary hydrogen atom in the Earth's gravitational field. However, I won't talk about that particular development, but rather about another one which the discovery of antimatter soon triggered.

Antimatter, then, was known to exist. At the beginning of the known history of the Universe, just after the Big Bang, the energy was so great that matter and radiation weren't distinguishable from one another. If that was indeed the case—and just about everyone admits that it is—it was only when the Universe had expanded enough for the concentration of energy to drop that matter could separate from light. But then, if there really was a symmetry between matter and antimatter, why isn't there as much antimatter as matter in the Universe? After all, that is what physicists observed in their giant particle accelerators: when they created matter, they created just as much antimatter.

The problem is that, if there were equal parts of matter and anti matter in the Universe, their meeting would mean that they would annihilate each other and subsequently unleash colossal amounts of energy. Let's calculate the order of magnitude together: the opportunity to use $E = mc^2$ is just too tempting. Let's imagine that a kilogram of hydrogen meets a kilogram of antihydrogen. That's very little on an astronomical scale, but the energy released would be on the order of $2mc^2$. With m = 1 kg, and c (the speed of light) = 300,000,000 m/s, that would yield 180 quadrillion joules, which is the typical amount produced by a group of six nuclear power plants over the course of a year.

If such a meeting between matter and antimatter took place in the Universe, there would be so much light produced that it would still be visible today. So why is there matter (the stuff we're made up of), but no antimatter in the Universe? This vexing question, which arose following the success of Dirac's equation, has not really been resolved, though we do know in what direction to pursue some answers.

It's possible that a fluctuation at the beginning of the Universe broke the symmetry at the start of the Big Bang, and that a slight excess of matter was enough for only matter to remain today. If I stand a pencil on its tip and let it go, it will fall to one side depending on its exact position at the moment I let go. Even if an ideal physicist prepared an ideal pencil and held it in an ideally balanced position, modern physics tells us that everything fluctuates, and so at the exact moment that the pencil was released, it would be impossible to be sure that it wasn't already leaning a bit in one direction. These questions of slight imbalance, or spontaneous symmetry-breaking, are at the heart of the progress physics has made in understanding state-changes such as the passage from a liquid to a solid state, and many others. But as for the Universe, the asymmetry between matter and antimatter seems to result from very profound causes.

For a long time, physicists believed that if they observed a mirror image of the world, all the laws would be the same, such that they would be unable to differentiate it from direct observation of the world without a mirror. We now know that this is false: mirror symmetry, also called "P parity," isn't preserved for some interactions. As long as we're dealing with electrical or magnetic interactions, or even gravitational ones (the attraction between masses), no problem. But there are two other interactions in nature, one called "strong," which is responsible for the cohesion of nuclear matter, and the other called "weak," which comes into play during some nuclear disintegration processes. In 1956, T. D. Lee and C. N. Yang argued that the weak interaction might not respect mirror symmetry, and

C. S. Wu demonstrated this symmetry-breaking by studying the emission of electrons from radioactive cobalt. It was thus proven that the weak interaction isn't the same in a mirror world: as physicists say, it "violates parity."

Given the current state of our knowledge, we think that a physical system obeys exactly the same laws if we make three transformations at once: not only a mirror symmetry P which transforms a left-handed chiral object (like a left hand) into its right-handed mirror image (a right hand), but also a temporal symmetry T which inverts the direction of time, and lastly, a charge symmetry C which flips matter and anti-matter. Do you know of Andrei Sakharov, the great physicist who, after inventing the Soviet H-bomb, heroically resisted his country's regime, earning him the Nobel Peace Prize in 1975? In 1965–67, when he was rallying the dissident intelligentsia, Sakharov suggested that changing the signs of the charges and parity (i.e., performing a CP transformation) should not leave the Universe unchanged. That means that the laws governing the Universe should change when the direction of time is reversed—at least those laws that have to do with the high energies involved at the beginning of the Universe's history.

Let's make this business about the direction of time more precise. Since Boltzmann,[6] we know that for an object on our scale, a "macroscopic system," time flows in the direction of increasing disorder: if I build a house of cards and I wait around for a while without touching it, it's highly likely that the cards will eventually fall to the ground in disarray. If I then hope for a gust of air to rebuild my house of cards, I'll be waiting for quite a while, since the probability of this happening is next to none. So the direction in which we perceive time to flow is determined by questions of order and statistics. But for a long time, we thought that it was different for particles. When we watch the film of a collision between two billiard balls, we're incapable of telling whether the film is being shown forwards or backwards. In the same way, we thought that for particles, the laws were independent of the direction in which time flowed. But that's false: we are now familiar with the disintegration processes of certain particles called "neutral K mesons," which violate the symmetry of time.

It's therefore possible that at the beginning of the Universe, the disintegration of certain particles gave birth to more matter than anti-matter, or that the lifetime of some particles was different from that of their anti-particles. Some physicists are actively looking for such processes, and are finding them. It was just discovered in 2004, in an experiment called by the name of "BaBar," that certain exotic particles called B mesons lived longer than their antiparticles. I certainly won't dare to claim that the mechanisms that violated the matter/antimatter symmetry at the beginning of the Universe are known, there are apparently several possibilities, nor that we understand why the density of matter is what we observe in the current Universe. However, that's one of the exciting directions for particle physics in the twenty-first century. There's a good chance that physicists will discover other particles, the behavior of which will make the "standard model," which describes the currently known particles, evolve even further. And then perhaps we will have a better idea of the origins of the asymmetry that allowed us—and stars, black holes, and of course, our dear cousins, the leeks, all made of matter as we are—to exist, such that we can ponder the mystery at all.

Perhaps this is why we are so fascinated with symmetrical inkblots, and why these images evoke such revealing, instinctive responses in us; they remind us somehow of our origins and of our relation to everything else in the Universe.

I Am Radioactive

Yes, I am radioactive, and for that matter so are you. The natural radioactivity of a human being is, as it happens, about 100 becquerels per kilogram (100 Bq/kg), and since I weigh 80 kg, my radioactivity is around 8,000 Bq. If you are a woman of 55 kg, your natural radioactivity is around 5,500 Bq.

Is 8,000 Bq a lot? How much is "1 bequerel" anyway? Am I in danger?

The bequerel is a very small unit that everyone should know about, given how much it's used with regard to radioactivity, and given how simple it is to understand. Radioactive elements are unstable and undergo reactions described as "nuclear" because they involve the elements' nuclei. During these reactions, which are spontaneous, the nuclei of "radioactive" atoms disintegrate to form other nuclei in a random manner and after a certain amount of time. In disintegrating, matter emits radiation, which could be composed of helium nuclei, electrons, very high energy light, or many other, more exotic things. The number of becquerels of a piece of matter is the number of its disintegrations per second. Now, every human body contains a little bit of potassium 40 and carbon 14.[7] These are two radioactive elements that are constantly emitting radiation in our bodies. The carbon 14 transforms slowly into nitrogen 14, basi-

26

cally by emitting an electron; potassium 40 transforms in an analogous way either into calcium 40 or argon 40. My body is therefore constantly being irradiated with 8,000 nuclear reactions every second. Happily, I'm a large collection of around 50 billion billion billion atoms, and 8,000, while perfectly measurable, is comparatively very little.

I'm tossing around these figures only because my passion for orders of magnitude grips me every time a newspaper gives figures without comparing them to others. That was the case in April 2001, when Le Monde echoed the accusations of an organization called the CRIIRAD: the "Independent Commission on Research and Information about Radioactivity," a private association whose label of "independent" means it's entirely unofficial. So, according to CRIIRAD, which had taken 3,000 samples in order to arrive at these alarming conclusions, the cloud from the Soviet reactor at Chernobyl had produced a serious radioactive contamination of cesium 137 when it passed over France in 1986; by the year 2000, that contamination had reached 5,000 becquerels per square meter (5,000 Bq/m^2) in some parts of Languedoc and 50,000 Bq/m^2 in Mercantour.

How many people were frightened by those numbers, with so many zeroes in them? What benefit was the CRIIRAD trying to extract from this manipulation of people's minds? If it had used a large unit such as the curie (Ci), which is the radioactivity of a gram of radium 226 (37 billion disintegrations per second), the radioactivity of those shallow ravines in Mercantour would only have been 1.4 millionths of a curie, and it would have created less of a sensation. I wish that more journalists would have the reflexes of physicists and, before writing their headlines, would put such numbers in perspective by first considering their orders of magnitude.

Five thousand becquerels in certain parts of the Languedoc is therefore perfectly measurable with a modern Geiger counter, but it's still less than the natural radioactivity of the person doing the measurement! And putting six people together in an elevator concentrates a radioactivity of 50,000 Bq within 1 m^2, which is as much as the alarming maximum registered on what some call the "leopard

27

spots" of the Mercantour. How many people realize that since granite is 100 times more radioactive than man, a granite pebble weighing 5 kg (barely a few centimeters of sidewalk) is enough, again, to reach the same 50,000 Bq considered by the CRIIRAD to be a catastrophe?

If the CRIIRAD accuses public officials of having hidden the existence of serious dangers related to the passing of the Chernobyl cloud,[8] it would be logical for them also to demand a ban on six-person elevators and granite sidewalks, as well as on climbing in the Alps and walking in Brittany. Wouldn't it?

I'm not pronuclear. On the contrary, I'd like it if we could avoid having to use nuclear power plants to produce the energy of the future. But is that possible? Probably not. This debate is of capital importance for the future of our way of life. To participate in the debate calmly, without being exploited by propaganda or pressure from one group or another and without just exchanging slogans, each citizen needs to learn just a few relatively simple notions about radioactivity.

I will come back to the future of energy production and consumption on Earth. But, for the moment, I'll settle for a few words about biology, archeology, and history—regarding the Earth and its daughter, the Moon, to show you that the reality of science sometimes outdoes science fiction.

At the time when Henri Becquerel and Pierre and Marie Curie discovered natural radioactivity, the phenomenon seemed so extraordinary that some people thought they would be able to cure anything by irradiating people. But Marie Curie died from leukemia, which was caused by the massive doses of radiation she absorbed while manipulating her radium all day. And then, of course, the Americans dropped two atomic bombs on Hiroshima and Nagasaki. Their motivation was probably to frighten the whole world, rather than to win a war that was already won. They succeeded perfectly, and it's for good reason that humanity is terrified by the enormous, destructive power of nuclear weapons. Between the absurd hopes of a century ago and the panicky fears of today, we must still try to reflect calmly and to consider what radioactivity really is: omnipresent in small amounts; having some very useful applications (including medical ones); a crucial source of energy, provided we manage to find a way to treat the waste from nuclear power plants (a problem which is far from solved); and the physical principle of a weapon of truly massive destructive power, since it's capable of blowing up the entire planet.

But it's not so simple to adopt a reasonable attitude towards radioactivity. In general, it isn't apparent when a body is radioactive. And besides, since we're all being naturally irradiated in a way that doesn't seem to interfere with living—residing in a region with lots of granite, like Brittany, doesn't seem to yield a greater risk of cancer than living elsewhere—it seems there is a threshold below which the risk is negligible and above which it is real. But does such a threshold really exist? If it does, can we define it precisely? That's a very interesting question, which no one seems to have answered as yet. We know that, in our cells, there are mechanisms for repairing the DNA molecules constituting our chromosomes. These mechanisms are necessary even in the absence of irradiation: there are errors of transcription and copying. Do cells have time to repair their DNA in the case of weak, rather than strong, radiation? That, I don't know. What makes me think that no one else really does either is that the legal thresholds of radiation are always calculated based on natural radioactivity: basically, if you undergo irradiation on par with natural radioactivity, the law decides that you aren't in danger. But if our exposure clearly goes over and above this natural radioactivity, the legislators decide to ring the alarm bell. While we wait for the radiobiologists to better understand how DNA is repaired in cells, this attitude of the legislators seems to me to be perfectly reasonable: we can't live as if radioactivity does not exist, so we need to regulate it and act cautiously while waiting for more information.

Some obsessively antinuclear people would like to see all irradiation banned, no matter what kind, but that's absurd; we can't eliminate natural radioactivity. But don't think that I'm ignoring the emotional aspect of these questions, and that I'm restricting myself to a narrowly scientific viewpoint which seeks to reduce them to simple numbers. I'm perfectly aware that the attitudes of scientists themselves are not purely rational. I suppose that Marie Curie, who manipulated radium with her bare hands (37 billion becquerels per gram, to be exact—the radioactivity of her samples would be enough to light up her laboratory on rue Lhomond at night!), couldn't be entirely unaware of the dangers that her research posed to her own health; however, she was solely interested in radiation therapy to treat the cancers of others.

During a brief stint at the "Conseil national des programmes," an organization under the aegis of the Ministry of Education, I argued that it was necessary for all high school students to have a minimal familiarity with the subject of radioactivity. If you were a French student enrolled in "terminal S" after 2001, you should therefore have learned that the discovery of carbon 14 shook the foundations of archaeology. It's a beautiful story, which unfolded in two acts. In 1934, the American physicist F. N. F. Kurie[9] noticed that exposing nitrogen to a beam of neutrons could produce radioactive carbon, ^{14}C. Then, in 1946, Willard Franck Libby had the idea that ^{14}C was constantly produced in the upper atmosphere of the Earth, where the nitrogen of the air is irradiated by the Sun. The Sun is an enormous ball of gas heated by nuclear fusion reactions; basically, it burns hydrogen and produces helium, and by doing this, it emits equally enormous quantities of different kinds of radiation, which we call "cosmic rays." Flying too long at high altitudes, let alone traveling in a spaceship or spending time on the Moon, is therefore dangerous to one's health. Life didn't develop on land until the atmosphere contained enough oxygen, and therefore enough ozone (a molecule containing three oxygen atoms), to protect living beings from the Sun.

It was verified that over the course of the successive reactions taking place in the upper atmosphere, protons were transformed into neutrons, and then captured by the nuclei of nitrogen atoms; stable nitrogen, ^{14}N, is thus transformed into radioactive carbon, ^{14}C. Once formed, this ^{14}C combines with oxygen to form the carbon dioxide gas $^{14}CO_2$, which is mixed with the rest of the atmosphere. Of course, in these molecules of carbon dioxide, the ^{14}C is slowly changed back into ^{14}N, but given the timespan over which this process has been operating, a stationary state has been achieved. The concentration of ^{14}C is stable throughout the atmosphere. But plants absorb CO_2, and we eat plants, so all living beings contain the same proportion of ^{14}C. As we've seen, that's one of the causes of our own radioactivity. On the other hand, once living beings die, their exchanges with the atmosphere stop, and since ^{14}C is transformed into ^{14}N, its concentration diminishes. By how much?

Every radioelement has a certain "half-life," as they say in nuclear physics, which is the time at the end of which the element has a one-half probability of having disintegrated. For ^{14}C, that's 5,700 years. That means that if one gram of ^{14}C emits 13.6 electrons per minute and we measure its radioactivity 5,700 years later, half of the ^{14}C will have been transformed into ^{14}N, and it will now only emit 6.8 electrons per minute. How, then, do we measure the age of a piece of pottery, or of a pile of ashes in a cave, or any other prehistoric object? All that's needed is to count the electrons coming out and weigh the object. This dating method was well worth a Nobel Prize, and Libby won it in 1960 (in chemistry, which proves the proximity of these two branches of natural science). The stakes involved in dating are so high that the method has been refined to respond to various questions, such as whether the stream of cosmic rays was constant throughout time. Today, the method allows us to date objects with remarkable precision, as long as their age is not much less than 5,700 years (the concentration of ^{14}C won't have changed enough) nor much greater (there won't be any ^{14}C left).

So not only is natural radioactivity a reality, but studying it has furnished archaeology with a method for dating objects, which was sadly lacking before. And that's not all: we can measure a wide variety of ages, thanks to a whole range of radioactive isotopes on Earth. We can also compare the different dating methods among themselves and verify that the ages determined are the same, regardless of which radioactive element is used.

For example, do you know how old the Earth is?

It's an old question, obviously. Allow me to pass quickly over the disputes concerning the interpretation of the Old Testament: give or take a few thousand years, all of these good people agreed that the Earth was 5,000 years old, which they also believed to be the age of mankind. However, not only is Lucy, the (female) hominid discovered in 1974 in Ethiopia, around 3 million years old, and not only is the hominid found in Chad in 2001 (Toumaï) around 7 million years old, but both of these ancestors of ours are extremely young compared to the Earth. *Homo sapiens* is even younger, despite being much older than biblical man: about 150,000 years old. The Earth itself is thirty thousand times older: $4.5 \pm .2$ billion years old. Mankind is therefore only a recent episode in the history of evolution, and the results of contemporary science should prompt a little modesty: mankind is only a recent consequence of the formation of the Earth, which is not the center of the Universe. The solar system itself is not, in fact, at the center of our galaxy, and happily so, since we would quickly be swallowed up by the black hole located there. As for our galaxy, the Milky Way, which I hope you gaze at during summer nights, like a great scarf of stars across the sky, it isn't at the center of the Universe either.

Quite a few scientists have tried to calculate the age of the Earth. In the 19th century, for example, Lord Kelvin calculated the time required for the Earth to cool to its present temperature, and determined that the figure was between 20 and 400 million years. But at the time, Kelvin wasn't able to include in his model the heat given off by natural radioactivity—mainly from the uranium everywhere under our feet: he didn't know that such radioactivity existed. In 1905 Ernest Rutherford suggested the figure of 140 million years by estimating the relationship between the uranium contained in the Earth and the helium produced when it disintegrates. But he hadn't thought of the fact that helium, a very light gas, is constantly escaping from the Earth. We see how much the interpretation of a measurement in physics depends on the model being used. Rutherford would nevertheless use helium nuclei, the "alpha" particles emitted by certain nuclear reactions, to demonstrate that the nucleus of an atom is much smaller than the atoms themselves. In the end, however, this idea of using radioactivity as a method of dating the Earth, by comparing the concentrations of elements produced by nuclear reactions, was a good one. In the 1950s, by comparing the quantities of rubidium and strontium contained in rocks, as well as potassium and argon, the age of the Earth was estimated at 3.5 billion years. We were thus approaching the truth which has today been established.

One day in 1994, while reading Jean Jacques's *Confessions of an Ordinary Chemist*, I came across a magnificent quote from Victor Hugo's *William Shakespeare*. Hugo, comparing art and science, wrote that "Science is the asymptote of truth. It approaches it ceaselessly, and never touches it. Besides that, it possesses every greatness. It has willpower, precision, enthusiasm, profound attention, penetration, finesse, strength, the patience of following-through, the permanent watch over the phenomena, the ardor for progress, and even touches of bravura." What an homage!

So, we were getting a little closer to the truth, when a new discovery took place, which raised questions about the most recent estimate of the age of the Earth: plate tectonics. The theory of plate tectonics describes the movements of the Earth's crust, in which different "plates" move about very slowly. They drift just a few centimeters per year—which seems rather slow, but which adds up over millions of years. After all, this so-called continental drift is responsible for India pushing into Asia, which thrust up the mighty Himalayas, and the process continues: the tragic earthquakes which shake that region, as well as others, prove it. And why do the plates move? Because the mantle underneath is actually a highly viscous, moving fluid. The center of the Earth has been cooling slowly since the time it was formed, despite the heat constantly being emitted from the disintegration of the uranium found there. There is, therefore, a flow of heat from the interior to the exterior. But anyone who has watched what happens in a pot of water being heated from below knows that a sufficiently intense flow of heat can create movement in a liquid: rolling movements appear caused by the hot water rising, since it is less dense than the cold water, and the descent of the water which has cooled at the surface, since it is, on the contrary, denser than the hot water. In the same way, the Earth's mantle is animated by large movements activated by the flow of heat between the center and the surface of the Earth. The motion pulls the continental plates along with it, often resulting in great quaking when the plates move against one another. Unfortunately for dating, all this shaking yields large exchanges of matter between the surface and the depths of the Earth, which has made the oldest rocks on the surface disappear. Hence, 3.5 billion years turns out to have been an underestimation of the real age of the Earth.

Then we landed on the Moon and brought back a few rocks. Their analysis suggested that the Moon was about 4 billion years old. That wasn't much different, but then the hypothesis was floated that everything in the whole solar system has the same age. This was confirmed by agreement between the analyses of several of the small meteorites which fall every day upon the Earth. After all that discussion, the final figure is a little higher and doesn't seem likely to change: 4.5 billion years.

On this question, as on many others, 20th century physics has made phenomenal progress, transforming our culture. It's no longer possible for us to situate ourselves in time and space the way our grandparents did. The age of the Earth is the same as that of the Sun and all the planets. But that's quite a bit less than the 13.7 billion years the Universe has existed. The Sun is thus a relatively young star, and I would be astonished if these estimates changed much—the Big Bang model seems to be in a period of refinement; it is no longer being seriously questioned. Now, if by chance I'm wrong, it would be very interesting for physicists: there would be a lot of work to do! Contrary to imposed dogmas, science is alive; scientific truths are meant to be discussed and disputed, eternally called into question.

Speaking of scientific truths, there is another one which is in the process of establishing itself: it concerns the formation of the Moon. Today we understand that stars form as a result of a collapse, due to the forces of gravitation: inside clouds of atoms and dust, matter attracts matter. But since not all of the cloud collapses, and the whole thing turns on its own axis, the star is usually encircled by a rotating disc, inside which more matter groups itself together, leading to the formation of planets. In the same way, discs (rotating rings) are collecting around some planets like Jupiter and Saturn, forming satellites. These satellites, though, are generally much smaller than the planet around which they revolve.

But now, the size of the Moon is not much different from that of the Earth. Could it be possible, then, that the Earth-Moon pair is a double planet, formed like a set of twins in the protoplanetary disc that surrounded the Sun? That seems impossible. For one thing, the composition of the Moon is different from Earth's: it resembles the surface of the Earth but lacks a heavy, iron-based core. That makes its mass 81 times less than the Earth's, even though its diameter (3,476 km) is more than a quarter of that of the Earth (12,756 km). Another hypothesis has been proposed, according to which the Earth, rotating rapidly on its axis, lost some of its surface after elongating. It doesn't seem, however, that the rotational speed of the Moon is compatible with this hypothesis. It has also been proposed that the Earth might have captured a foreign object which was passing by and kept it in orbit; but then, if this object was already part of the same solar system, why didn't it contain the same kind of metallic core as the Earth? Now, in our day of computers and complex calculations, the idea proposed by Hartmann and Davis in 1975 has apparently been accepted: the Moon is a result of a titanic collision between the newly formed Earth and an object the size of Mars (around 6,800 km in diameter). According to calculations done in 2004 by Robin Canup, an American scientist from Boulder (Colorado), if that planet collided with the Earth in a glancing way rather than full-on, it could have removed some of the Earth's surface and its central core could have combined with the Earth's. Then, the enormous, and now volatile, cloud of light elements around the Earth collected, or accreted, to form the Moon. It even seems that this accretion process only took a few years! Finally, the Moon would have gradually lost energy, and in so doing would have drifted progressively farther away, until reaching its current distance: around 50 times the radius of the Earth.

By way of all of these digressions, I've gone from radioactivity to the history of the Earth. While we're here, I may as well add two more things: one on dinosaurs and one on extraterrestrials. Let's start with the dinosaurs. All twenty-first century schoolchildren know that there was once a wide variety of dinosaurs and that they have practically all disappeared. When was that? Once again, we've learned through dating methods that the last dinosaurs died exactly 65 million years ago, just at the end of the Secondary Period (the Cretaceous) and the beginning of the Tertiary. But why did such a brutal event take place? Here is another example of an especially active controversy which has challenged quite a few scientists in their search for a good explanation. Two main hypotheses were in conflict, each ardently defended by opposing schools of geophysicists. One group maintained that an enormous volcanic eruption took place, which would have unleashed so much dust into the atmosphere that a considerable cooling would have followed. It would have been night for years. If a nuclear war were to take place, a "nuclear winter" of the same type could follow. Whatever we may imagine about our future, though, the winter of 65 million years ago would have been followed by a global warming, and the whole process would have so perturbed the ecosystem of the planet that large numbers of animal species would have become extinct and numerous others would have appeared.

The second hypothesis resembles the first except that the catastrophic origin of the winter is different: according to this hypothesis, it was triggered by the impact of a meteorite at least 10 km in diameter. The energy released would have been equivalent to 100 million megatons of TNT. This second hypothesis is supported by the fact that the geological layers contemporaneous with the extinction of the dinosaurs contain exceptional concentrations of iridium—a rare element which also happened to be used, as I learned at school, to make a standard meter kept at the Breteuil pavillion in Sèvres.[10] The concentration of iridium is 10,000 times greater in meteorites than on the Earth's surface. The probable site of impact was also identified, in the north of the Yucatan in Mexico: an enormous crater several hundred kilometers in diameter. The controversy hasn't stopped, however, since volcanic lava containing more iridium than the Earth's surface has also been discovered, and scientists have identified a volcano in India whose activity must have combined with the impact of the meteorite to force the planet into a period of mass extinction. It seems, however, that the results from the most recent numerical models are more supportive of the second hypothesis than the first: there is too much iridium for it to have been the product of this lone Indian volcano.

Life on Earth, then, almost died 65 million years ago. Speaking of which, are there other Earths? I mean, of course, are there other planets with life? Perhaps you think that I'm drifting into science fiction, but I'm not in the least. As of February 2008, close to 300 "extrasolar" planets (also called "exoplanets") have been discovered around stars near our own solar system. To detect their presence, we measure things like the luminosity of stars. When a planet passes in front of a star, the light we receive from it drops a bit; and when the partial eclipse is complete, the star's light returns to its earlier value. We don't yet have images of these foreign planets, but it won't be long! The telescopes which are currently being constructed are like giant pairs of eyes, which are becoming increasingly farsighted.. It thus seems possible that by 2025 we could obtain images of exoplanets, the resolution of which would be high enough to distinguish continents and clouds. The number of known exoplanets in 2025 will no doubt be considerable. The probability, therefore, is high that at least one among them will resemble the Earth. Will we discover life there, and even extraterrestrials?

If it were only a question of finding a planet with about the same dimensions as the Earth and roughly the same distance from a star similar to our Sun, the probability of finding life elsewhere would be rather high. But that's just it: it seems that these conditions aren't enough. Let's suppose we're looking for a life-form with a similar chemical composition to one we might find on Earth, that is, one which originally developed in liquid water. It is generally agreed that that would require, just like on Earth, the temperature to be not be too far from $0\,°C = 273\,K$, so that there can be liquid water, ice, and water vapor, that is, water in its three principal states. But something would have to prevent this exo-Earth from losing its water by evaporation, and it seems that the stability of the orientation of the axis of the Earth is important for that. If the Earth rotated on an axis pointing towards the Sun rather than being tilted 23 degrees 27 minutes in relation to the plane of the ecliptic, it would be day for six months on one side and night on the other, then the opposite for the following six months. Apparently that would have more than sufficed for the Earth to lose its water. Even if the Earth's axis oscillated a little too much, and thus occasionally approached this extreme orientation, the water would evaporate. Even though the fluctuations of the axis of the Earth are minor, they have important consequences for the climate: they are what produce cold periods (ice ages), followed by warm (interglacial) ones every few tens of thousands of years.

So we are starting to think that to conserve water on a planet's surface requires a relatively stable axis of rotation which doesn't point towards the Sun, like that of the Earth. But the Earth's axis is stable only because it's stabilized by the Moon: the Moon is not a small satellite—it's size is comparable with that of the Earth, and as you know from the tides, the movements of both are linked. So, it seems that in order to detect life on a planet's surface, we may have to discover an exo-Earth with an exo-Moon, and that makes things complicated. Since, as we've just seen, the Moon might be the result of a slightly angled collision between a Mars-sized planet with the Earth a short time after their formation, the probability of finding an exo-Earth/exo-Moon couple resulting from a similar collision near a star is very low. But still, it's worth searching.

It is likely that one of the first tests to detect a habitable, Earth-like planet will concern the presence of ozone. Life under water produced oxygen, which gradually escaped from the water and was then partially transformed into ozone by the Sun's radiation. After that, the ozone protected the surface of the Earth from the Sun's ultraviolet radiation and allowed life to emerge from the water. But that took time! Life out of water began only 600 million years ago, despite the fact that it began in water about 3.5 billion years ago, six times earlier. In order to find the signature or evidence of a kind of life resembling what we know, we would have to find an exoplanet that is not too young. So if science fiction novels have given you the desire to meet evolved beings who know enough physics to send us radio messages, then you're asking too much. In effect, *Homo sapiens* is only a recent evolutionary adventure, between 100,000 and 200,000 years old. Modern man was only able to develop when, between two ice ages and after hundreds of millions of years of evolution out of water and onto land, climatic conditions became favorable enough. But even 200,000 years is nothing on the scale of geological time. It's not at all certain that mankind will survive a possible ice age 20,000 years from now, not to mention all the climatic disorder which excessive oil consumers are currently triggering. So, to find exo-people, we would need a very happy coincidence, one which would allow us to meet this exo-Earth at about the same stage of evolution. In short, after having believed for some time in the rather high probability of our meeting extraterrestrials, I've returned to a rather more prudent pessimism.

But it is worth looking, and wouldn't it be exciting if we found a truly different life-form? Our conception of the world would be, well, shaken up a bit. In the meantime, we'll just have to wait and see what the year 2025 brings, and anticipate that—just as it was a surprise to learn so much by pulling at one thread, the study of radioactivity—we may continue to be surprised again and again as we continue our search to reveal the history of our own Earth, the Solar System, and how our home planet compares to other possible homes for life many light years away.

4

Einstein, the Flute, and Quantum Dew

1995. Seventy years had already passed of investigating whether this pesky business of a particular kind of condensation thought up by Einstein was a dream, a mistake, or a reality. What was involved was the experimental verification of a new state of matter, which Einstein had predicted in 1924, based on the work of a young Indian theoretician, Satiendra Nath Bose. Did it really exist? In 1995, like marathon runners finishing their race in a sprint, two American teams simultaneously discovered that what the Bose-Einstein equations had been predicting for decades was, in fact, true. Physicists were walking tall. The predictions of their science, surprising as they might have seemed at first, and in spite of any doubts which had arisen from such a long wait, had turned out to be right. Their science was powerful, since it was capable of dependable predictions. Sodium in Massachusetts behaved like rubidium in Colorado, just as Einstein and Bose had predicted. Thousands of researchers around the world were going to leap into the study of this phenomenon and pick apart its most minute aspects. In response to such a success, the academy in Stockholm wouldn't wait long to reward the three sprinters in this long-distance race with a shared gold medal for physics: the Nobel Prize of 2001.

And so, in December 1995, faced with the urgent need to com-

Figure 2. From the cover of *Science* 270: December 22, 1995. Reprinted with permission from Steve Keller and AAAS.

municate the news of this successful discovery, the American magazine *Science* came out with a squadron of blue atoms marching in step on the cover. It wasn't so much the color blue that bothered me. After all, why not? Sure, sodium is better known for its yellow color—throw a pinch of salt into a fire and in the flame you'll see flashes of the same yellow that you find in the harsh illumination of some underground passages and highways. As for rubidium, an alkaline like sodium—only larger and lower down on the periodic table—it would be more of a red, after its name, which suggests rubies. But after all, this midnight blue no doubt symbolized the great coldness of these atoms, and pretty well suited the scientific truth under construction.

No, what bothered me was actually that march—those orderly troops. True, I had bad memories of my own experiences with military marches. Before May 1968, the hierarchy at the École Polytechnique had little patience for the antics of its rebel students, and my deviance had cost me fifteen days in prison and gotten me barred from marching with my fellow students on the Champs Élysées.[11] But regardless of these youthful memories, I saw in that march of atoms a basic error of interpretation as to what the recently discovered "Bose-Einstein" condensation was. The order of the actual condensate seemed radically different to me from the regimentation depicted on the magazine cover.

The artist at *Science* hadn't just dreamt up this march all by him or herself, though: the military analogy had been hanging around in the public scientific discourse for a long time. Nevertheless, in becoming famous, this image threatened to distort the understanding of the discovery. I intended to denounce it firmly.

It has to be said that the role of images in physics is surprising. We physicists aren't satisfied with merely observing nature; we claim to understand what we observe. You might think that that consists of writing equations to describe measurements which have been performed and in general, predicting what the results of future measurements will be—if we change certain conditions of an experiment, for example. But I know few physicists who rest content with that. It's as though their display of perfect rationality doesn't prevent them from in fact preferring, at the end of the day, a kind of intuition rooted in analogies—even approximate ones. I admit that I myself fight in vain, however doggedly, against all sorts of unfounded myths, unjustified beliefs, dogmas handed down from on high, various obscurantisms. I am not satisfied simply with the result of a calculation; I also need a representation to nourish my imagination—words and images which support my conviction. Proof is rigorous, but deep conviction doesn't come to me without the construction of a link to my daily experience.

Are we sure we really understand how a copper wire, for example, conducts electricity, when we recite Ohm's law, which we learned in school? How do we come to understand that electrical resistance depends on the cross-section and the length of the wire? Isn't it easy to tell ourselves that electrical current is like the water flowing through a pipe, that the difference in potential is like the difference in height between the top and bottom of the pipe, and that the current is therefore more intense if the height differential is larger, if the diameter of the pipe is greater? And, instead of learning Kirchhoff's current law, which describes the distribution of electrical current between three branches of a circuit, isn't it easier just to imagine that the current in a river is the sum of the currents of its tributaries, since water flows everywhere without accumulating anywhere?

Sometimes these images are deceptive because the analogies are false; indeed, the study of certain very fine electrical wires proved at the end of the twentieth century that the electrical resistance we know so well from Ohm's law isn't *always* proportional to the length of the wire. Must we continue to hide that from middle school students? That's a pedagogical question which doesn't seem at all simple to me; to teach science, must we reduce it to a handful of its provisional truths? It's a fact, then, that we need images to fully comprehend; the equations are not enough for us. We are rational scientists, and yet we need imagination to understand. I wonder how far my colleagues who are mathematicians go down this path, they who champion rigor but support their proofs with so many drawings in the margins of their calculations.

The magazine *Science* had thus taken an image which was widespread among physicists in order to communicate, in a more tangible way, what "Einstein-Bose condensation" might be. So what is it about that blue brigade, then, that makes me so indignant, even today? And can I manage to describe and explain this "condensation" invented by Einstein without equations? Is it possible to explain in plain English the nature of a discovery made in the arena of quantum physics—a branch of physics renowned for being so completely removed from our daily experience?

Let's see.

Around a hundred years ago, quantum physics was born out of the need to explain the structure of the atom—I'll come back to that in the next chapter. Bohr, Einstein, Heisenberg, Schrödinger, Bose, Fermi, Dirac, de Broglie, and many others overturned our understanding of what matter and waves are, especially light, by discovering that everything was both wave and particle at the same time. Since quantum physics isn't immediately obvious, to say the least, avoiding all mathematical formalism will force me to use analogies to clarify the paradoxes and apparent contradictions of this surprising theory. In fact, it's precisely because comparisons to daily life make quantum physics seem so astonishing that, for a hundred years, some physicists have strenuously tried to prove that it is false. Let's say right away that these efforts have thus far been in vain: quantum physics has resisted every attack mounted against it, since being organized into a coherent framework, usually called the "Copenhagen interpretation," by Niels Bohr.

Let's come back to Bose-Einstein condensation. It is completely different from the condensation of dew—when water vapor forms droplets on the kitchen windows on a winter morning—which I'll call "classical" condensation. Rather, it is a grouping of atoms "in a single quantum state." What, then, is a "state" in quantum physics?

First analogy: let's imagine a vibrating string, like one we might find on a violin. The violinist drags her bow along the string to create a vibration. She gets a sound out of it, amplified by the body of the violin, which has a certain pitch. This pitch depends on the length and tension of the string. Unless she changes this length by

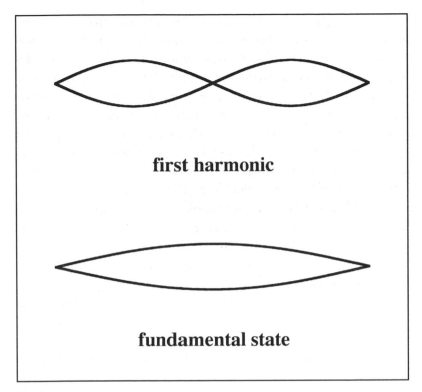

first harmonic

fundamental state

Figure 3

moving her finger on the neck of the violin, or she changes the tension by turning the little key at the scroll, the violinist playing on this string will always get the same sound—that is, a sonic vibration of the same pitch, or in scientific language, the same "frequency."[12] The string always vibrates in its fundamental "state," the one with the lowest frequency. However, violinists know well that it's possible to produce what is called a "harmonic," by lightly brushing the string with a finger properly placed so as to force it into a different mode of vibration. Instead of a vibration which is zero at the two ends and intense at the center, that is, with what are called two "nodes" and an "anti-node" for the vibration, it is possible to produce a third node at the center. The frequency of this mode of vibration is twice that of the fundamental, and the string produces a sound one octave higher.

On a flute, the column of air enclosed in the pipe is what resonates; it can't be seen, but it is just like a violin string. Just by making slight changes to the position of his lips, but without moving his fingers and without changing the length of the resonating air column, a good flautist can play six or seven different notes, the frequencies of which are successive multiples (by two, three, four, and so on) of a fundamental frequency. If, for example, he closes all the stops by pressing down with all of his fingers, he can play a low C, which has a frequency is about 262 Hz; but he can also play the next C, which has a frequency double to that, then some other notes near the G immediately above (triple the frequency), another C (quadruple), an E (five times higher), another C (six times) and maybe even a B-flat (seven times).

The sonic frequencies corresponding to the different vibrational states of the flute—of the column of air within the flute, that is—cannot therefore be just any number. The frequency of a flute is not a continuous variable. We may say, rather, that it is a "quantized" variable, just as in quantum physics. That means that it can assume only specific values, connected to each other by a simple relationship: in this case, by multiples of two, three, four, and so on of an ultimate, indivisible quantity: what physicists call a "quantum." These possible values constitute a series that we call "discrete," to distinguish it from an infinite, continuous series.

Well then! It's the same for a quantum particle. An electron in a small box, if it were classical, i.e., obeyed the laws of classical physics, could move at any speed and thus have any kinetic energy; if you play squash or baseball, you know perfectly well that you can throw the ball, which is a classical object, at whatever speed you like (or are capable of). However, a quantum electron is also a wave, and if it is enclosed in a box, it will resonate just as a violin string resonates between its fixed endpoints, or like the air column in a flute. For that reason, the resonance frequency of the electronic wave in the box, and therefore its energy, cannot take just any value. The electron has access only to specific modes of resonance which are called "quantum states," and which have respective energies that are successive multiples of a quantum, corresponding to a fundamental frequency. The energy of the electron in this "quantum box" is therefore "quantized," since it is a resonating wave.

I sense skepticism on your part. These are not the abstract fantasies of an unverifiable theory, however. We are able to make these boxes today. They are only a few nanometers in size, a few billionths of a meter, comparable to the wavelengths of the electrons we inject into them. When we build these boxes, we take care to make sure that they contain no defects or obstacles likely to absorb the energy of the electrons resonating inside. When the electrons change from one state of vibration to another, they emit light whose color depends on the difference in frequency between the states, which is fixed by the dimensions of the box. The experimenters can thus adjust the color of the light emitted by these quantum boxes however they wish. Those boxes also represent the future of microscopic light sources called "integrated lasers."

Let's come back to the principles of quantum physics. Particles are waves, then, and their accessible states in a quantum box are quantized. And an atom, of course, is a sort of box containing a nucleus and imprisoned electrons. That's why, for example, every sodium atom emits the same yellow light, and oxygen molecules all emit the same blue. But you know, don't you, that the sky is blue for a different reason, related to light being scattered by the atmosphere? And besides, the blue of the sky isn't the same everywhere.

Now let's imagine that we put two electrons, or even more, in a box. What will happen? Might the electrons all behave in the same way or, rather, must each behave differently? On a ladder, each rung of which represents a different quantum state, will the electrons place themselves all together on the same step, or will they distribute themselves evenly, with one at most on each step?

It has been shown that there are two types of quantum particle: those which can all adopt the same behavior, and those which cannot. A widely used metaphor is that Bose particles, called "bosons," are herdish, whereas Fermi-Dirac particles, called "fermions," are individualistic. Electrons are fermions and so can't sit on the same rung all together, which is to say that they cannot occupy the same quantum state. These differences in collective behavior, which I've been trying to present in imagistic language, are in fact regulated very rigorously by precise statistical laws. Are there really only two types of possible behavior for groups of quantum particles? That is a question that is close to the cutting edge of physics, since particles, or rather groups of particles, have just been discovered which are neither fermions nor bosons—but I don't think I'll be able to explain these other particles simply, so let's ignore these recent oddities and assume, for the present discussion, that there are only two possibilities.

What makes a quantum particle a boson or a fermion? Simply put, it's a question of internal rotation which leads to a magnetization called a "spin," the value of which is either an integer multiple $(0, 1, 2, 3, \ldots)$ of a certain quantum, or a half-multiple $(1/2, 3/2, 5/2, \ldots)$. Electrons, protons, and neutrons are individualistic fermions of spin $1/2$. On the other hand, photons, which are quanta of light, are herdish bosons. But a pair of fermions together can constitute a boson, since when their spins are added together, $1/2 + 1/2 = 1$. (Excuse me if I'm being overly thorough.) Thus, the hydrogen atom, which consists of a proton and an electron, is a boson, just like an atom of helium 4 (two protons, two neutrons, and two electrons); but its light isotope helium 3 (two protons, a single neutron, and two electrons) is a fermion. These facts yield some remarkable possibilities. For example, when the electrons of a metal pair off with each other, the metal becomes superconductive, its electrical resistance disappears, and enormous electrical currents can pass through it without resistance. It's thanks to this phenomenon that we are able to produce the intense magnetic fields required by magnetic resonance scanners, that is, the modern medical imaging technique used in hospitals. I will come back to this several times later on.

Boson or fermion, similar or different collective behavior—these are not abstract curiosities but verified realities, the applications of which are beneficial to our health. We have almost arrived at the condensation predicted by Einstein and Bose. The only missing pieces are a few points concerning the effects of temperature on all of this.

Temperature is a measure of the agitation of matter. The temperature of a gas at rest is simply proportional to its kinetic energy, that is, to the square of the speed of its atoms or molecules. The only subtlety in this simple law is that the temperature involved is what is called "absolute" temperature, which is measured in degrees Kelvin, and whose symbol is K, rather than the ordinary temperature of household thermometers which is generally measured in degrees Celsius (°C). But the Ks are easily derived from the Cs by adding 273.15. So water freezes at 0 °C = 273.15 K, and boils at 100 °C = 373.15 K. Of course, converting degrees Fahrenheit into degrees Kelvin is a little more difficult.

To give an idea of the relationship between temperature and thermal agitation, we may say, by way of example, that in the air of a temperate month of June, which might have a temperature of 27 °C = 300 K, the oxygen molecules are moving at an average speed of 200 m/s = 720 km/h. That's almost the speed of a passenger jet. At 100 K, or −173 °C, the average speed of the oxygen molecules is only about 114 m/s, close to the speed of a fast train like the French TGV, and the oxygen isn't far from liquefying. And if the temperature approaches what is called "absolute" zero (0 Kelvin, or −273.15 °C), the thermal agitation stops and in principle, matter stops moving. It then liquefies and freezes—that is, it crystallizes (except in the case of glasses and liquid helium). To observe Bose-Einstein condensation in 1995, it was necessary to cool atoms below a microkelvin, a temperature at which the speed of these atoms is very slow, on the order of a centimeter per second.

The atoms or molecules of a gas are classical when hot, and quantum when cold. Why? With nothing to contain them, the particles are also waves, but waves whose extension in space depends on the temperature. They are, in fact, more accurately described as wave packets, the typical length of which was calculated by Louis de Broglie, and which is inversely proportional to the thermal velocity just mentioned. Two cases are possible, then. First case: the waves associated with the particles overlap because their extension is at least equal to their average distance of separation. So the particles are affected by the wavelike character of their neighbors, and they become indistinguishable from each other—the gas is quantum, whether the particles are bosons or fermions. Second case: the spatial extension of the wave packets is slight, the particles are far away from one another compared to the de Broglie wavelength, they are not affected by the waves of their neighbors, and so are classical, well-separated, and identifiable, not quantum—and thus neither bosons nor fermions. Wolfgang Ketterle has illustrated this on his website in order to explain the Bose-Einstein condensation, which won him the Nobel Prize.

It's at cold temperatures that the difference between the two types of particles is apparent. If you cool bosons, they not only will all become progressively indistinguishable from one another, but they will also group themselves together in the same fundamental state, i.e., they will all be on the same rung of my imaginary ladder—the lowest one. On the other hand, if you cool fermions, they will also be indiscernible but will collectively occupy a number of states equal to the number of particles, i.e., they will form a cloud spread over as many rungs as there are fermions. "Bose-Einstein condensation" involves bosons (as you may have guessed). It is produced once the temperature gets low enough for a high proportion of the atoms to have fallen into the lowest-energy state. If you've followed me this far, I imagine that you find the phenomenon a bit abstract. I agree with you. That's why, like my fellow physicists, I need to represent to myself what a Bose-Einstein "condensate"—this assembly of atoms in the same state—might look like.

Does it look like troops marching in step? Those who prefer this image are thinking of the fact that the atoms move together at the same speed if they are in the same state. Their movement is ordered. This uniform behavior makes some people think of a march. But if that's right, they are representing an order of movement by an order of position: they represent their atoms as firmly positioned at the nodes of a moving network. Such a condensate, then, is really only a crystal in motion. But in fact, that's not the right image at all. As I tried to explain above, this order appears when the atomic waves overlap. That means that the atoms are no longer discernible from one another. If we persist in imagining them as particles, they are

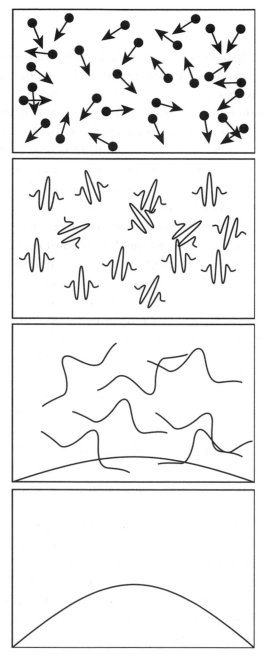

Figure 4. Wolfgang Ketterle's rendering of Einstein-Bose condensation. At high temperatures, the atoms of a gas are agitated particles moving incoherently in all directions. As the temperature drops, their wavelike character is manifested, and the atoms become wave packets that spread progressively outwards, overlapping each other, then "condense" in the form of a collective wave below a critical temperature TC. At absolute zero, the atoms have lost all individuality and form a giant wave of matter which can resonate in a box like a violin string between its endpoints. From top to bottom: high temperature $(T >> TC)$, low temperature; $T = T_C, T = 0$.

constantly exchanging places; if we adopt the wave interpretation, instead, they are superimposed waves which occupy all the space available to the gas.

Anyone who has been in a military march knows that he would run the risk of being arrested if he got it into his head to exchange places with his neighbor. It would definitely be noticed! Of course it would, since in a company marching in step, each soldier is distinguishable from each of his neighbors. Even if two twins marched side by side, we could pick one out as the one who was on the right as opposed to on the left, or locate him as the one in the front, there, in the middle of the first row. In a quantum condensate, however, there are no more individual atoms, just a big wave of macroscopic matter. Ketterle's drawing is less pretty than that of the marching soldiers, certainly, but it is more accurate scientifically. The one in *Science* confuses identical behavior with ordered positioning. It ignores a troubling paradox of quantum physics: at a low enough temperature, atoms become spread-out waves which overlap and are indistinguishable from one another.

The 1995 discovery was far from the first proof of the wavelike character of particles. We've known for a century how to use electrons to obtain interference patterns which exactly resemble those obtained by using two agitators in a basin in order to create waves that meet. These waves combine if they are in phase, but annihilate each other if they are out of phase—that is, if the water is rising in one wave while dropping in the other. Today, we can make larger and larger particles interfere in this way; for example, molecules of C_{60} whose 60 atoms of carbon form a sort of spherical grillwork, a network analogous to the pattern on a soccer ball. We are trying to demonstrate such quantum properties for larger and larger objects, in order to understand where—if anywhere—the boundary is in terms of size between quantum matter and classical matter.

Speaking of soccer balls, I'll end this discussion of quantum physics with a mathematical curiosity. Imagine a grid formed of regular hexagons. It is supple and easy to roll into the form of a cylinder. But if you want to fold this hexagonal grid into a sphere, you'll be forced to twist its hexagons. The great mathematician Euler proved that to manipulate easily a hexagonal mesh grid into a spherical shape, we need only to replace 12 regular hexagons with 12 regular pentagons. A soccer ball effectively has 12 pentagons in the middle of a certain number of hexagons, as does a molecule of C_{60}. We may also insert heptagons, but then Euler's theorem says that the number of hexagons is always arbitrary, while the number of pentagons, N5, minus the number of heptagons, N7, must equal 12. In general, soccer balls have no heptagons and therefore 12 pentagons, but some architects (like Buckminster Fuller, who uses grid-like structures) know they are constrained by Euler's theorem if they wish to insert heptagons. Finally, this theorem generalizes: the number N5-N7 only depends on the kind of shape involved, that is, on the number of holes; it is thus 12 for spheres but 0 for cylinders or for a torus, a class of mathematical shapes which includes air tubes for bicycles and doughnuts. I leave you to figure out what it would be for a shape with two holes, a sort of pretzel.

5

Is My Table Quantum?

`\ \ \ \ \ \ \ \ \ \ \ \ \ \ \ \ \ \ \ \ \ \ \ \ \ \ \ \ \ \ \ \ \ \ \ \ \ \ \`

One spring morning I went to my neighbors' place on the fifth floor to talk about gardening, but instead, their daughter preferred to talk about physics! She lacked neither for curiosity nor good sense, and so I tried to intrigue her and shed some light on aspects of modern physics at the same time. We ended up here:

"... I have a quantum table? No, now you're making fun of me, Sébastien!"

Of course, as an everyday object for eating on, a table has nothing to do with quantum theory. However, suppose I asked you how much it weighs? Or, what amounts to the same thing, what the mass is per unit of volume of the wood composing it—its density, let us say for short. Wood floats on water, so its density must be a little less than that of water, which is one gram per cubic centimeter ($1 \, \text{g/cm}^3$). In fact, all massive materials have a density on the order of $1 \, \text{g/cm}^3$. But why would $1 \, \text{cm}^3$ of solid or liquid matter weigh roughly $1 \, \text{g}$? The density of a gas can be very low because its atoms are very far apart from one another, but in a solid, the atoms are stacked together compactly. Basically, to ask about the density of a solid amounts to asking how much volume an atom occupies. But the answer to that question has to do with quantum mechanics. What, then, is the volume of an atom?

All solids have similar densities because all atoms themselves have comparable size; their diameters are all on the order of 0.3 nanometers.[13] To line up a billion atoms one after another, whatever kinds of atoms they are, like pearls on a necklace, would require about 30 centimeters of (very thin) string. Speaking of atoms, it has been a long time since their status was upgraded from that of a hypothesis to that of a verified reality. In 1979, Gerd Binnig and Heinrich Rohrer, two physicists in the IBM laboratories near Zurich, invented a new type of microscope, the "scanning tunneling microscope," which yields not only images of atoms one by one, and not only captures the difference between an atom of silicon and an atom of oxygen, but which also lets one manipulate these atoms, picking one up and moving it a bit further along, so that one can play at writing with them like a child writing with pebbles on a beach.

Let's think about the hydrogen atom, which is the simplest one. It's also the smallest and lightest, and so we'll find that its density is a bit less than $1\,g/cm^3$. Actually, if we dramatically cool gaseous hydrogen, it passes to a liquid state at about 20 Kelvin ($-253\ °C$), then to a solid state at around 14 Kelvin ($-259\ °C$). Solid hydrogen has a density of $0.08\,g/cm^3$, but for approximate calculations, that isn't very different from $1g/cm^3$. An atom of hydrogen has a proton at the center, the nucleus, with an electron orbiting around it. The proton has a positive electric charge and its mass is rather large; as for the electron, it has the same charge, only negative, and about 2,000 times less mass. One might call it a miniature Sun with a little planet revolving around it—but look out! That analogy was thrown out no less than a century ago (although it is still taught in many high school physics courses). In effect, like any object whose movement obeys Newtonian classical mechanics, a planet moves along a trajectory: at any instant we can measure where it is and what its speed is, just like a falling apple. But this no longer holds when we get to quantum particles: *quantum mechanics abandoned the notion of trajectory*, which brings us straight to one of its bizarre aspects, the Uncertainty Principle. This principle was formulated by Heisenberg and stipulates that one cannot precisely know both where a particle is and the speed at which it is traveling; if we know one well, we know the other poorly, and vice versa. Since some have a tendency to extrapolate this principle beyond physics in risky ways, it's a good time to demystify it a little.

In fact, the uncertainty principle isn't so difficult to understand if we accept the basic postulate of quantum physics: a particle is also a wave. Enclosing a wave in a box means knowing where it is—that is, within a box—but that also means losing the possibility of knowing the direction and speed of its motion inside that box. In the reverse manner, to properly measure the speed of a wave, it's necessary to let it propagate freely in space and thus no longer know very well where it is. The electron of the hydrogen atom is confined to the interior of the volume of the atom, a fraction of a cubic nanometer. Within that little volume, it is so excited that it moves about in all directions, and the smaller the volume, the greater the excitement. It's a bit like the air resonating inside a flute: the shorter the pipe, the more often the sound wave goes back and forth between the two extremes, the greater the resonance frequency, and thus the higher the sound pitch. As I mentioned in the last chapter, today we make lasers with such quantum boxes, and the light they emit has a frequency which gets higher, and thus an energy which increases, as the dimensions of the box get smaller. The uncertainty principle expresses this conflict between position and speed in a rigorous manner: if we confine a quantum particle in a region of space, the smaller the region, the greater the average speed of the particle, and hence also its kinetic energy.

The typical dimensions of an atom are thus a compromise: the proton attracts the electron because their electrical charges are opposite. This electrical force pulls the electron closer to the proton, to minimize the volume of the atom in order to minimize its electrical energy. But if this volume were very small, the quantum excitement of the electron due to Heisenberg's uncertainty principle would be very high. Its speed would endlessly fluctuate between high values in every direction, which would give it a high average kinetic energy. Since the stable state of a physical system is the state of lowest energy, the electron finds itself at a distance from the proton, which is a compromise between the electrical attraction exerted by the proton and a sort of repulsion due to quantum excitement. In theory, we might imagine that we must also take the gravitational attraction between the proton and the electron into account, since the two particles both have mass, and so they attract each other in the same way that the Sun attracts the planets. But gravitational attraction is much weaker than electrostatic attraction, which is why solar systems are much larger than atoms.

I need to take a few precautions with my comparisons, however. A sound wave is an oscillation of pressure and density that propagates itself; an electromagnetic wave like light is an oscillation of electrical and magnetic fields. But the quantum wave, which represents a particle, is only an oscillation of the *probability* that the particle has of finding itself at a particular place. That's certainly bizarre, but don't be too surprised: *everything about quantum physics is bizarre*.[14] Regardless, the probabilistic character of quantum mechanics has an important consequence: within the atom, the electron is a wave which is spread out in this little region of space. It doesn't revolve like a planet; it is, in a way, diffused by its probabilistic nature, like a sort of random cloud.

How can we find the precise value of the compromise that the electron makes? It requires some real calculation, taking into account the exact charge of the electron and the amplitude of the Heisenberg quantum excitement, which is proportional to Planck's constant, so it gets a bit difficult. Let's simply recall the result given earlier: 0.3 nanometers in diameter. Taking into account the mass of the hydrogen atom, which, expressed in grams, equals the inverse of Avogadro's number—$1/(6 \times 10^{23})$—we get the right result for density. Thus, the density of solid matter is about equal to that of the atoms which constitute it, which is a consequence of a fundamental principle of quantum physics, Heisenberg's uncertainty principle. Only with the advent of quantum theory, in the 1920s, could scientists understand the most basic facts about matter.

It's time for a word about the color of atoms, and another about clocks. If you remember the different states of the vibration of air in the tube of a flute (last chapter), you'll understand that an electron can be in a fundamental state of low energy, but also perhaps in different excited states which correspond to higher-frequency modes of resonance. A particular energy fixed by quantum mechanics corresponds to each state. Where do the colors of matter come from? With a few exceptions,[15] they are due to the fact that when an electron changes state, it absorbs or emits light. Since the possible quantum states of electrons are entirely determined by the nature of the atoms to which they belong (their number of protons, neutrons, and electrons), all the atoms of a single kind have exactly the same color. By jumping from one state to another, their electrons absorb or emit exactly the same grain of light—the same photon. When a Parisian throws a pinch of salt into the fire, its flame has exactly the same yellow color as the flame in front of a Japanese person, when he or she does the same thing in Kyoto: sodium atoms have the same color in Paris as in Kyoto (and it's not because the two cities are sister cities). It's true everywhere else, as well, including other planets.

This property of atoms—that they are identical everywhere—is harnessed in the construction of atomic clocks, which set the standard whenever a high degree of precision is necessary. The color is related to the frequency of light being absorbed or emitted, and atomic clocks work by measuring the inverse of this frequency, which is time; they display the exact same time everywhere, as long as they use the same type of atom: usually it's cesium, and the optical properties of all cesium atoms are the same everywhere, according to quantum mechanics. In contrast, from recent observations of the large number of planetary systems in the Universe, we have found that the periods of rotation of these planets around their respective suns are different. There is no reason for them to have any particular values, since they are classical objects, not quantum particles.

The conversation with my young neighbor took a sudden turn:

"But now that I think about it, if an atom is a small nucleus surrounded by a cloud, it's completely different from a marble or a pearl on a necklace; could you then push two atoms together, so their clouds either compress each other or blend together, and so that the nuclei could come within a much smaller distance of each other than the diameter of each atom's cloud?"

An interesting question, but no, the clouds can not compress or blend with each other. Remember that electrons are fermions (see previous chapter)—that is, individualistic particles which cannot occupy the same quantum state together. This statistical property, which is also called "Pauli's exclusion principle,"[16] justifies my analogy with necklaces. In effect, two electrons may occupy the same quantum state as long as they belong to two different atoms, well-separated from one another. If their clouds overlap each other to form a single cloud, the two electrons can no longer be in the same state. Of course, molecules are associations of atoms, but that's another story; let's leave that aside and stick to general principles. Two identical atoms repel each other at short distances because electrons are fermions, and their clouds don't mix. At short distances, two identical atoms are really like two hard spheres, mutually impenetrable.

Allow me an aside: the exclusion principle is what gives chemistry its richness. Since each element corresponds to a different number of electrons, 1 for hydrogen, 2 for helium, 3 for lithium, 6 for carbon, 7 for nitrogen, et cetera, the way in which these electrons are arranged is different for each element, and a considerable diversity of electrical, magnetic, and chemical properties result in nature. Let's try a metaphor: each species of atom has its own character, and so prefers to relate with some atoms more than others (chemists speak of "affinities"). Thus, the complexity of chemistry is also a result of the individualism of electrons. And, let us note in passing that, just like the color of atoms, this diversity of chemistry is the same everywhere—whether it's in your stomach, in an appealing perfume, or in interstellar clouds.

"So, is my table quantum?"

—We cannot explain all of its properties without appealing to quantum physics, which is omnipresent—though it is hidden for some people, and more obvious for others; that's a matter of taste.

"But then, if there's something quantum about my table, and if quantum physics says that matter and waves are the same, shouldn't we be able to propagate 'table-waves'?"

—Ah, now your enthusiasm has taken you a bit far afield, but that question makes me think of the story of Schrödinger's cat; and if I manage to tell it to you, we should learn enough of quantum mechanics to examine one of the other current frontiers of knowledge.

Schrödinger, one of the discoverers of quantum physics, continually challenged his own and others' understanding of the subject. Schrödinger's cat was both dead and alive *at the same time*. Yes, of course, that's obviously impossible; nothing with which we are familiar can be at the same time one thing and its complete opposite. So this dead/living cat can't exist. It is, however, a paradoxical conclusion reached by Erwin Schrödinger, and in a way that deserves consideration.

Take a spinning top. When you let it loose, you can make it spin to the right or to the left. And, like the cat, the top can't spin right and left at the same time. But quantum particles can.

Many charged particles effectively rotate around an axis, and rotating electrical charges produce magnetic fields. So, many particles have a little magnet whose orientation depends on the direction of rotation of the charges. This little magnet is quantized: the orientation of the magnet's field is called the "spin," and it can be directed upwards (in which case we'll say that it has "spin up"), or downwards (we'll say that it has "spin down").

So, a classical top cannot spin both clockwise and counterclockwise, just as a compass can't indicate north and south at the same time, and yet a quantum particle can be in a spin-up state and a spin-down state at the same time, which we call superimposed states. In this situation, if we measure the spin—that is, its rotational state—we get "up" with a certain probability and the opposite state with a complementary probability. Once again, as long as we allow that particles are waves, we can understand this a little better. Light, for example, is an electromagnetic wave: a magnetic wave and an electrical wave which oscillate continuously in space as the wave propagates. Consider a water wave: the water rises and falls as the wave advances. In the same way, light is constituted by an electrical field E and a magnetic field B, which are perpendicular to one

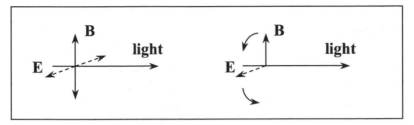

Figure 5

another as well as to the direction in which the light is traveling, and which oscillate.

But these fields can oscillate in several ways: they can oscillate vertically or horizontally, or even rotate around the direction of propagation. But now, vertical oscillation is the sum of a rotational motion in one direction with another in the reverse direction. And horizontal oscillation is another sum of two rotational movements opposite to one another. A wave can therefore rotate to the right and to the left at the same time—it's a matter of polarization, which opticians make use of to create polarized sunglasses to attenuate bright reflections from car hoods, water, et cetera.

Schrödinger imagined the following experiment. We place a particle in a box. A detector measures its magnetization (its spin). In a neighboring cage, Schrödinger imagines a cat. If the detector measures spin up, it triggers a mechanism which kills the cat. If, on the contrary, the detector measures spin down, the cat stays alive. Schrödinger then imagines that his particle is in a superposition of two opposite states of spin. He deduces from this that the cat is both dead and alive at the same time!

Impossible, of course, but why? Subtlety is required here, but it's very important to clarify a particular point. In order for a physical system to be in one state superimposed on its opposite, it must be a quantum system which can be described as a wave. It doesn't matter whether it's a wave resonating in a box or a wave propagating freely, but it must be a wave (with a frequency and an angular phase which tell us where the physical magnitude is in each cycle) not an incoherent noise. Now we're getting to the heart of the difference between a classical system (basically, all the objects of our familiar environment) and a quantum object (particles for the most part, with a few remarkable exceptions). It's the notion of *coherence*. To represent waves to myself, I like to think of waves on the surface of water. If I place in the water a little agitator which oscillates regularly, it emits waves on the surface of the water—circular, concentric, which move away from the source of the agitation at a certain speed. I have created a circular, coherent wave. If, on the contrary, I agitate the water in all directions and in a random manner, by making it boil in a pot for example, its surface will be highly agitated, but I will be unable to identify any particular wave, with a well-defined distance between the propagating wavefronts. A quantum system is a coherent wave, contrary to a non-quantum system, that is, a classical one. For example, a laser is a source of coherent quantum light, whereas an ordinary lamp is an incoherent classical source.

How can we differentiate the two systems, classical and incoherent versus quantum and coherent? By making interference patterns. If I use a laser to illuminate a screen in which I've made two slits, light passes through each slit and arrives on the other side. What passes through is a superposition, or a sum if you like, of what has passed through each slit. This superposition is not uniform but rather has a shape consisting of dark and bright bands. This is an interference pattern, since in certain places the amplitudes of the oscillations coming from each slit are added together (the two electrical fields are both directed upwards, for example), while in other places the fields arrive with opposite signs and cancel each other out. You can make interference patterns in water: if you use two agitators which emit circular waves, there is a region where these waves cross. Or you can try—why not?—with your fingers in the bathtub; you will see that the intersection of the two circular waves has a particular structure, with zones that don't move and other zones that move—the equivalents of dark and light bands in luminous interference patterns. But if you illuminate the same two slits with an ordinary lamp instead of a laser, you won't see an interference pattern: what comes across will be uniform light. In this way

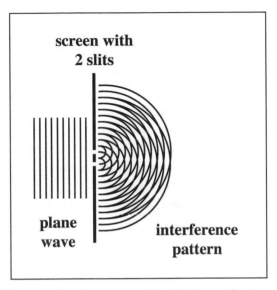

Figure 6. A plane wave that hits a screen pierced with two slits is transmitted to the other side as two waves that interfere with each other, producing an alternation of bright and dark zones.

we can distinguish coherence from incoherence: it amounts to asking, in essence, whether the waves are well or badly defined.

Imagine a particle which is propagating in empty space. If quantum mechanics is true, it's a free wave, without interactions with anything, and it should therefore go through both slits at the same time and produce an interference pattern behind the screen.

"Hang on! If I send an electron against a plate with two slits, the electron goes through the right and the left slits at the same time?"

"Yes! I know that goes against your intuition, but it's true, it's been verified. Just as an electron can spin to the right and to the left at the same time, *it can go through the right and left slits at the same time*. Unbelievable or not, this experiment is now classic. We can detect the transmission of a beam of electrons across two slits; we find zones with electrons and zones without, like bright and dark bands. We can even send electrons one by one, and we find that they always arrive in certain zones and not in others, proof that they go through both slits at the same time, just like the waves they are."

Before going any farther into the disturbing paradoxes of quantum physics, let's determine the fate of the poor cat. In order to be one thing and its opposite at the same time, a physical system must be coherent; now the electron in Schrödinger's box might be coherent, but not the aggregate of the electron + cat. So that is why the cat cannot be alive and dead at the same time. And why isn't a cat a quantum system? Where is the boundary between the quantum and the classical, if there is one at all? This question is fundamental and at the cutting edge of research today; the answer to it will also lead to potentially revolutionary applications.

I've given the impression that the answer has something to do with size. Particles are small, and they can be quantum; cats and tables are large, and they are classical. That's actually pretty close to the truth, as it happens. Let's think about our wave again. As long as it propagates freely, no problem; we see clearly that the water rises and falls periodically in time and space. If the wave is resonating in a basin, it becomes a stationary wave, like the sound in a flute, and it's still a well-defined wave. Being reflected against the side doesn't destroy the coherence of a wave, as long as there is no loss of energy. It's a bit like a billiard ball: its motion after reflection against a cushion is predictable on the basis of its prior motion. But if the cushions absorb energy—if there is an exchange of energy between a particle (a wave, a billiard ball . . .) and its environment, then everything changes, and the coherence of the wave is lost. If a laser beam is absorbed by a piece of black paper and re-emitted, it's no longer the same wave—it's no longer even the same color (or frequency). In this way, an interaction with the environment can destroy the coherence of a wave. The larger the system, the more it interacts with its surroundings, and the less it's able to preserve the coherence of the wave which would represent it if it were quantum. A table is strongly coupled with what surrounds it: it rests on the ground, it absorbs and reemits light. A cat meows or purrs and thus emits sound, as well as heat. In sum, the coupling of a macroscopic object with its environment is so strong that it's impossible for it to preserve the coherence of a wave, even for the shortest imaginable time. So don't try to send cats, apples, or tables through double slits;

it's useless, and there is no hope of observing interference patterns on the other side.

But then, again, where is the boundary between the classical world and the quantum world? It's a matter of time, size, and coupling: the longer you want the coherence to last, the more it's necessary to limit the number of interactions between the object and its environment, and therefore its size. To come back to the crucial test of interference patterns across a double slit: in Vienna in 2001, Anton Zeilinger managed to measure interference patterns with the molecules of C_{60} mentioned previously—the spherical grids of 60 carbon atoms, which look like miniature soccer balls. The experiment was a tour de force, and we don't yet know whether it's possible to successfully conduct this experiment with objects larger than a C_{60} molecule. There are, however, two remarkable exceptions, namely, superfluids and superconductors: they are waves of matter which are both quantum *and* macroscopic.

Do you remember the Bose condensation? In the last chapter, we saw that a gas of sufficiently cold atoms could reach a state in which all the atoms, having become indistinguishable from each other, group themselves into a big wave of collective matter. In 1995, researchers at the University of Colorado at Boulder and MIT discovered that this quantum phenomenon was indeed as Einstein had predicted seventy years before. Even though this discovery was truly a breakthrough, scientists had been persuaded for a long time that the same phenomenon was responsible for both the astonishing properties of liquid helium at low temperature, and the properties of numerous metals, which become "superconductors" again at low temperatures.

Above 2.17 Kelvin ($-271\,°C$), liquid helium is a classical liquid, light and normal: if you launch a flow of liquid helium into a container, it will eventually stop moving—since it's slightly viscous, it rubs against the sides, and friction will slow it down. But below this temperature, which is pretty cold but easily obtained in a laboratory, the viscosity disappears. Some experiments showed that if you launch a flow of sufficiently cold liquid helium into an extremely thin but porous glass ring, the flow won't stop—it flows for months without measurably slowing down. When the Russian physicist Piotr Kapitza observed this property of exceptional flow, he coined a term to name and convey the phenomenon: "superfluidity." This was at the end of 1937, and the story of that discovery will be the subject of the next chapter. We'll see that Kapitza didn't choose the word by accident but rather with reference to a phenomenon he considered to be similar: "superconductivity."

Superconductivity had been discovered in 1911 by H. Kamerlingh Onnes, at Leiden. The Dutch researcher had measured the electrical resistance of mercury, which is solid and metallic at cold temperatures, and discovered that it disappeared below about 4 Kelvin. The presence of electrical resistance in a metal comes from the fact that electric current is nothing other than a flow of charged particles, generally electrons, and that this flow involves the loss of energy: the electron gas is mobile between the ions of the metal but rubs against them while moving. Kapitza thought that helium stopped rubbing against the sides of containers below a certain temperature exactly like an electron gas in certain metals. He was right, but it took a long time for scientists to be persuaded. The theory of superconductivity was established in 1954 by a trio known by the nickname of BCS—John Bardeen, Leon Cooper, and Robert Schrieffer—which of course won them a Nobel Prize (in 1972). BCS figured out that the electrons of a metal could associate pairwise in order to become bosons, and then condense in Einstein's sense of the term. They also figured out the origin of the attraction that pairs the electrons. It resembles, to return to a metaphor I used earlier, the situation that happens between two people on a soft mattress: each body (each electron) forms a hollow in the mattress (in the crystalline network of the metal), into which the other will tend to fall. But it's not possible for me to explain all that here. Scientists have therefore been convinced that liquid helium is superfluid for the same reason that metals are superconducting, even if a few rebellious minds persist in doubting: Bose-Einstein condensation does take place. The difficulty which prevented physicists from admitting this fact for a long time was that in a liquid, atoms are strongly mutually interactive, whereas Einstein's calculation concerned a gas with no interactions. A lot of work was required to generalize Einstein's results and to be persuaded that the same phenomenon was at work. In fact, even today, the precise calculation of the structure of liquid helium is considered rather difficult. In contrast, for a gas of cold atoms condensed in Einstein's sense, everything is calculable with precision, and the theory was thus verified beyond doubt.

I've gone on this long digression into superfluids and superconductors in order to point out a remarkable exception to the rule: quantum objects are not always microscopic. Weren't we just saying that the best proof of whether something is quantum or classical is an interference experiment? Exactly! It turns out that it is possible to make interference patterns with a superconductive current traveling along by two different paths, or with a flow of superfluid helium, which passes through two slits or holes. We really do see constructive and destructive interference—the equivalents of dark and white bands with laser light, which proves that this matter is a coherent wave.

How is that possible? For complex reasons, these superconducting or superfluid currents can't exchange energy with their environment; their coherence remains intact despite the large number of atoms involved and the large number of neighboring atoms. I just wanted to point out this exception, for the sake of full disclosure. I should also say that, in my own laboratory, I've seen superfluid helium almost every day for the last twenty years, and it continues to surprise me even now.

Quantum physics is difficult even for professional physicists. I want to finish with a few remarks about quantum computers, which you may have heard about in the news lately.

If you have accepted that there is some evidence of quantum mechanics at work in a simple table, then it doesn't seem like too much of a stretch to assume that, in a computer, there must be quite a lot of quantum mechanics taking place! Well, of course, it's all a question of perspective. We mustn't confuse science with the technical objects that science has enabled us to construct. In the case of computers, it's certainly true that there are many transistors, and that the transistor was invented as a direct application of the quantum theory of semiconductors. But the logic of the computer is classical. By that I mean that the calculations are performed by the operations of circuits, and there is nothing indeterminate about that. No logical port on your computer is both open and closed at the same time, for example. No unit of memory contains both the value 0 and the value 1. But certain calculations are now being made according to an entirely different, quantum logic. For the moment they are elementary, though they may progress further. Their logic uses a superposition of states to perform calculations in parallel on several elements at the same time. What's the point? The hope is to solve a number of so-called "complex" problems much more quickly, which could, among other things, revolutionize the encoding of secret messages.

"Complexity," in common language and when used by many scientists, is a rather poorly defined term. Often a problem is called complex when we just don't know how to solve it easily. However, in certain cases, complexity has a real mathematical definition which is rigorous and not very hard to understand.

A complex problem, in the sense of applied mathematics, is like that of the traveling salesman—someone who has to visit a certain number (N) of cities and wants to find the shortest route. Of course, if there are just two cities to visit, there's no problem. If $N = 3$, it's still rather simple: there aren't many different possible routes, and the traveler will easily find the shortest one. But if N is very large, the number of possible routes is considerable; furthermore, even if we happen to chance upon a reasonably short route, one that is in fact even shorter could turn out to be an extremely different route on a map, such that we wouldn't be able to find it by just tweaking the other one a little bit. The modern traveler will program his computer to solve this problem and will notice that it takes it a certain amount of time to find the solution. How does this time vary as a function of the number of cities? Is it proportional to N? That seems surely false. Does it vary like the square (N^2) or the cube (N^3) of N?

It has been shown that, in fact, the time it takes to solve this problem increases exponentially, that is, faster than any polynomial function like N, N^2, N^3, et cetera. This increase is so rapid that if the number of cities to visit is truly large, the time required to find the right route would be equal to or greater than the age of the Universe. Mathematicians say, therefore, that this problem is "NP-complex," which means "nonpolynomial," impossible to solve if the number of parameters is too large.

Another example of a complex problem is that of the decomposition into prime factors of a large number with many digits. As long as the number of digits is small, solving the problem goes pretty quickly. For example, finding that 15 equals 3×5, or even that 210 equals $2 \times 3 \times 5 \times 7$, does not pose much of a problem, either for you or for a computer. However, if it's a number with 15 digits, it will be necessary to try dividing it by all the prime numbers one after another, which takes a considerable amount of time. For encoding secret messages, we use decomposition into prime numbers of numbers that are large enough so that a computer would not be able to find the factors in a human lifetime.

What does all this have to do with quantum computers? The point is that, in reality, complexity is not an intrinsic property of a problem. Everything depends on the manner in which the computer looks for the solution. It's therefore a property of the problem and the computer taken together. If we invented a type of revolutionary computer which would be able to calculate more efficiently than all the current computers, certain complex problems could become more simple to solve. This goal is what motivates much of the research on quantum calculation at the moment. Theoretically, it seems possible to use a large number of quantum-coupled elements (specialists say they are "entangled") to perform many simultaneous calculations over all the elements at the same time, and that considerably reduce the time necessary for calculation.

Will the quantum computer be invented tomorrow? Will classical computers one day be replaced by quantum computers, thereby revolutionizing calculation and reducing the complexity of problems we face? The technical difficulties involved are daunting. The main one resembles the business about Schrödinger's cat: the computer is a large macroscopic object. Calculating in parallel over two or three paired quantum particles is possible, and has already been done; some researchers have triumphantly announced that they successfully found that 15 equals 3 times 5. Others smiled. As interesting as the new principle of calculation is, putting it to work still seems out of reach at the moment: as the number of quantum elements which need to be coupled increases, the amount of time over which the necessary quantum coherence resists the environment shrinks, and scientists don't, to my knowledge, know how to get around this major difficulty at the moment. But they're looking for a way forward. And the intersection between information theory and quantum mechanics is clearly a fertile area of research, at least on paper. Perhaps researchers will even find something other than what they're currently looking for, which often happens. In any case, quantum physics has been surprising us for a century, and may have even more surprises in store.

The Power of Words

One day in March 2003, Aiko Oshima and Vincent Ciccone, two students at the School of Decorative Arts, came to see me. They wanted me to write some text for them: more precisely, to pick 20 words related to my work and then to write definitions to these words. By collecting several contributions of this sort from various people, they hoped to generate a creative, collective description of life on this hilltop in the Latin Quarter in Paris, la montagne Sainte-Geneviève, where there are numerous institutes for advanced study and research and where art and science rub shoulders, even if the agents of each don't always realize it. The two students were planning to juxtapose my text with their own works of art and exhibit them together. I liked their idea. I went along with it, though some of my definitions were aimed at exciting curiosity rather than really explaining anything. This is more or less what I wrote for them:

Garden: A space with plants. Rare in the Latin Quarter. The garden separating the Decorative Arts from the laboratories of the Ecole Normale Superieure and from the Institut Curie is an experimental botanical garden. One day the physicists of the ENS, at the behest of their director, Edouard Brézin, decided to roll up their sleeves and do some landscaping in the garden, to make it a blossoming place where we could picnic in summer.

Green: The color of my blackboard. The others are white. To my knowledge, in this neighborhood, only Pierre Gilles de Gennes, at the *College de France*, has a red one.

Helium: The lightest and simplest of the noble gases. An element that is often studied in physics, because it has no inherent chemistry. It's used to fill balloons for children, so that, according to Archimedes's Principle, they'll rise upwards. It is also the coldest and purest of all liquids, which makes it an exceptional candidate for the study of the properties of matter, whether classical or quantum. To approach absolute zero, it's an indispensable coolant, and without it, medical imaging wouldn't be what it is.

Coffee: I remember a tiny dark room at the Kapitza Institute in Moscow, where the local "coffee club" would meet every day for endless exchanges on the latest advances in physics. In June of 1984, I joined this club for one (symbolic) ruble. More science was done in that place than any other room in the building, a building, incidentally, which Stalin had built, taking advantage of Mendeleev's hundredth birthday to repatriate Piotr Kapitza by force. Kapitza was a student of Rutherford at Cambridge and a future Nobel Prize winner for his contributions to the discovery of superfluidity. At 24 rue Lhomond in Paris, Albert Libchaber also tried to add coffee to the general weekly seminar, since there's really no place where colleagues run into each other easily in the building. At some point, space had been saved, but to the detriment of natural collaboration among the people working there.

Diagonals: Cesium is the heaviest of the alkalines. In the periodic table, it is all the way on the bottom left, while helium is all the way on the top right. In 1991, Jacques Treiner and his three American colleagues Cheng, Cole, and Saam discovered that this diagonal separation prevents liquid helium from spreading on solid cesium. Another Jacques, this one the head of our design office, is a dedicated cyclist; he already has several "diagonals" to his credit, like the road from Brest to Nice.

Bicycle: A means of transportation, particularly suited to rue d'Ulm, which is often jammed with traffic due to ambulances from the Institut Curie, funerals at the Lebanese church, and garbage trucks. Jealous of my mobility, drivers stuck in the traffic jams honk at me when I pass them on my bicycle. I think of Marie Curie, who planted these beautiful plane trees almost a century ago, on my left.

Cold: Below zero degrees Celsius (0 °C), it's cold on rue Lhomond. Cold is essential to my research, and my dilution refrigerator goes down below −273 °C. Zero on the scale of absolute temperature (0 Kelvin = −273.15 °C) is an impassable boundary, where the thermal agitation of matter stops. But, there is still quantum agitation, which prevents liquid helium from freezing (unless you compress it).

Cosmonauts: In May 1968, thanks to a serendipitous (mis)stroke of paint, demonstrators renamed the convent on rue Lhomond: the "missionaries of the Holy Spirit" became the "cosmonauts of the subconscious." Regardless of whether NASA thinks it's our mission to send humans up into the heavens, it's a mistake to send people into space. It's expensive, dangerous, and robots do better than people at the only task for which it's really worth going up there in the first place, namely observing the Earth and the cosmos around us.

Vacuum: In the twentieth century, physicists learned that the vacuum is not empty. It's the site of fluctuations which are responsible for the spontaneous emission of light by atoms. Despite this fact, when you pump the air out of a hermetically sealed container, you remove practically all of the atoms it initially had, which changes its properties, in particular, the transmission of heat across the emptied space—a vacuum is a good thermal insulator.

Leaks: The nightmare of low-temperature physicists. By filling up the empty spaces (which serve as thermal insulators) with gas, leaks make all cooling impossible. The worst kind of leaks, known as "superleaks," let only superfluid helium through—since below 2 absolute degrees (−271 °C), superfluid helium is a quantum liquid with no viscosity, and it can leap through the smallest hole as if it were a giant open tunnel.

Email: An electronic message. By extension, the communication network itself. All the researchers throughout the world are now permanently interconnected by way of their computers. Via the World Wide Web, they exchange a lot of science, much friendship, and sometimes more. This network was invented in 1989 by Tim Berners-Lee, a researcher at CERN, in Geneva. For a long time, scientists have been one big family spread over the world.

I hope that these words had the power to successfully convey certain aspects of my daily life to my young neighbors. If I worked on elementary particles, I would have made perhaps better use of the metaphorical power of certain words from the language of science; the "charm" of particles is, for example, a rigorous concept, not a simple question of aesthetics. On that topic, I have no objection to scientists who use metaphors to help represent concepts. Thus, when I talk about "gregarious bosons," it's in order to convey the tendency of these quantum particles all to behave in the same way. But when psychoanalysts like Jacques Lacan or Julia Kristeva misuse the words of quantum physics or mathematical logic in order to invest themselves with a false authority, I protest along with Alan Sokal and Jean Bricmont.[17]

In any case, words don't just have the power of association. They are, of course, the basis of thought. But when a scientist invents a word, does he also invent a concept? Could it be that he is only facilitating the communication of an idea by picking it out with a single word? Should we give all of the credit for the discovery to the scientist who coins the word, or rather simply recognize that he is better at communication than others? This question often gets asked in a highly competitive milieu such as ours, where researchers are frequently evaluated, especially via the process of conferring various types of honors—invitations to conferences, national and international prizes, for example. Our microcosm rewards individuals, who then appear to be the sole authors of discoveries which, in most cases, are really the results of collective labor. The research community is highly sensitive to these honors, and legitimately so. But there's a tendency to remember only the winners, and when a single winner has been named, in spite of all the scruples that the jury may bring to bear in making its choice—especially for the

Nobel Prize—it's sometimes necessary to add a little nuance to our judgment of the players.

I think that's the case with superfluidity. The example is particularly close to my heart, since superfluidity is a quantum property I've studied, but also because the history of its discovery in 1937–38 illustrates several aspects of scientific research and sheds light on the troubled period in which it was made, as well as on the impact that the invention of a word can have.

"Superfluidity," then, is the property certain fluids have of flowing without friction—that is, as if their viscosity were zero—below a certain temperature. Its experimental discovery is often attributed to Piotr Kapitza, who won the Nobel Prize in 1978, and its interpretation to Lev Landau, an immense figure in Soviet theoretical physics, who won the Nobel Prize sixteen years earlier in 1962. I made a few enemies (though also some friends) by daring to contest this pair of attributions in the *Journal of Low Temperature Physics*.[18] Let's go back over the complex history of the affair.

On January 8, 1938, the British journal *Nature* published two articles side by side. The first, on page 74, had been sent on December 3, 1937, by P. Kapitza of the Institute for Physical Problems, Moscow: "Viscosity of liquid helium below the lambda-point." The second, on page 75, was sent on December 22, that is, nineteen days later, by J. F. Allen and A. D. Misener of the Royal Mond Laboratory, Cambridge: "Flow of liquid helium II."

The "lambda point" is the temperature ($T_\lambda = 2.17$ degrees Kelvin $= -271°C$) at which, in 1927 at Leiden, near Amsterdam, Willem Keesom discovered an anomaly in the physics of this very cold liquid: at that temperature, helium is particularly difficult to heat. In effect, what physicists call its "specific heat" is at its highest at that temperature. The form of the curve that represents that quantity as a function of temperature resembles the Greek letter lambda, whence the name Keesom chose. Afterwards, he called the liquid above T_λ "helium I," and the liquid below T_λ "helium II." At that point, he had understood that despite being composed of little round atoms with no chemical properties, the liquid he was dealing with presented two different liquid states—a rather surprising discovery.

75

Events gathered speed in April 1937 at Cambridge, when J. F. Allen, R. Peierls, and M. Z. Uddin found that helium II conducts heat remarkably well. In Moscow, their competitor Kapitza immediately began trying to find out why. If its viscosity were low, might the movements of the liquid be especially intense, and contribute to the transmission of heat in that way? He tried, therefore, to measure the viscosity by studying the liquid's flow through a slit less than one micron thick, between two polished discs pressed against one another. He found that above T_λ, helium I barely flowed, whereas below it, helium II flowed very easily. In his *Nature* article, even though he gave no quantitative results of measurements, he affirmed that the difference in pressure between the entrance and the exit from the slit was proportional to the square of the velocity, and then deduced from this that *the flow was turbulent*. Pursuing this line of reasoning (which seems rather confused to me), he then explained that *if the flow were not turbulent*, that is, if it were laminar, the viscosity of the liquid would have to be extremely low—less than a billionth of a poise,[19] that is, ten million times less than water. In conclusion, Kapitza finished with two lines which became a landmark: "By analogy with superconductors, the helium below the lambda point enters a special state which might be called superfluid." Thus he invented the word "superfluid."

As I mentioned briefly in the last chapter, a metal's electrical resistance disappears and it becomes superconducting when its electrons flow without friction. Liquid helium was another example of flow without friction, and Kapitza was right to compare these two phenomena even though the relationship between them would be made clear only decades later.

To sum up: Kapitza's article is astonishingly thin and partly false, but it contains two lines of genius at the end. Let's compare it with the article that was published alongside it.

The article by the British scientists begins: "A survey of the various properties of liquid helium II has prompted us to investigate its viscosity more carefully. One of us had previously deduced an upper limit of 10^{-5} cgs units for the viscosity of helium II by measuring the damping of an oscillating cylinder. We had reached the same conclusion as Kapitza in the letter above; namely that, due to the high Reynolds number involved, the measurements probably represent non-laminar flow." So, between the 3rd and 22nd of December 1937, Allen and Misener had learned of the arrival of their competitor Kapitza's article. I found out how. But first let's situate Allen and Kapitza in their own turbulent historical context.

Kapitza was a Soviet researcher who had earned his PhD at Cambridge—the University of Newton and many others—under Rutherford, at the Cavendish Laboratory founded there by Maxwell. Thanks to a generous grant from Ludwig Mond, the Royal Society had built the Mond laboratory for him. There, Kapitza had invented and built a new type of helium liquefier. He also performed dangerous experiments there with very high magnetic fields; the laboratory was therefore built lengthwise, to protect the director of the laboratory from any eventual explosions at the other end of the building.

He was in the habit of returning to Moscow every summer to see his family, but in 1934 the centenary of Mendeleev was being celebrated at Leningrad, so Kapitza had two reasons to return. That's when Stalin prevented him from leaving. The great dictator needed the great scientist to help achieve the oxygen production required for industrial development, and later to help with the construction of the Soviet bomb, which would lead to Kapitza's disgrace. In 1934, Stalin wanted first and foremost to put his kidnapped scientist to work. So he built the Institute of Physical Problems in Moscow for him, a sort of a large house in a park, which we know today as the Kapitza Institute. After the kidnapping, Rutherford intervened in order to send Kapitza his equipment, on two conditions: first, that the Soviets would purchase the material, and second, that Cambridge got to keep the precious helium liquefier constructed by Kapitza.

Helium, I should make clear, was starting to acquire a special place in modern physics. Even before realizing that quantum physical phenomena could manifest themselves on a macroscopic scale, scientists had understood that helium was indispensable in the race for low temperatures. They also knew that near absolute zero

(0 degrees Kelvin), the thermal agitation of matter stops, and that very delicate interactions between the constituents of matter become measurable, just like a landscape becomes visible when a fog lifts.

But the number of laboratories where it was known how to make liquid helium could be counted on one hand: Leiden, Toronto, Oxford, Cambridge, and Kharkov. In order to build another liquefier and restart his scientific activity, Kapitza had also managed to bring one of his students to Moscow for three years—David Schönberg, the future director of the Mond Laboratory—as well as the two technicians who had helped him to liquefy helium at Cambridge, Laurman and Pearson. We can see how much Stalin needed Kapitza; so much so that when arbitrary arrests were rampant, Kapitza was able to secure the release of a young theoretician named Fok, as well as that of Landau himself. Landau, a future leader of theoretical physics in the USSR, had done his doctoral studies in Leningrad and then a postdoctoral stint with Niels Bohr in Copenhagen, before getting his first position at Kharkov (Ukraine) and then being invited to Moscow to work with Kapitza. But in March 1938, the arrogant young genius was arrested and imprisoned for opposing the regime. To get him out of prison, Kapitza had to offer his own freedom to Molotov as collateral for Landau's, but it still took him a year to manage it (in April 1939). So just before confronting Stalin and Molotov, Kapitza (who didn't lack for guts!) had sent his article in to *Nature*.

The importance of publications for all scientific researchers can't be overstated. To be published in a scientific journal was already at that time the proof of the production of a piece of knowledge. The best journals have been international for a long time, and each has an editorial review board that notes an article's date of arrival, for obvious reasons of maintaining a system of priority, and keeps the article secret for the time required to critically evaluate it. That stage consists in checking whether the content is coherent and rigorous and whether it holds up in the face of critique by two anonymous referees. This review phase is important and delicate, and we all respect it scrupulously, regardless of the disputes that may pop up from time to time. Acceptance for publication in a peer-reviewed international journal is what confers the first gauge of scientific quality on our work, what gives our results the first element of truth. And of course, the one who publishes first is the discoverer, and the second looks like he simply reproduced the results of the first. Moreover, in principle, every publication should lay out all the details required for another researcher to reproduce the work in question— that's a rule, whether it's a measurement being made or an equation being solved. A scientific result must be reproducible. Nevertheless, is it possible to claim that someone has reproduced someone else's work when his research was done more or less at the same time? That's one of the problems we scientists face.

So, Kapitza had sent his article in to *Nature*. But we can imagine his impatience and his desire to be published quickly, as well as the difficulties in communicating between Moscow and England in 1937. That's why he asked John Cockcroft, who had been his colleague while working under Rutherford, to correct the proofs in his stead. Cockcroft had taken over as director of the Mond Laboratory after Kapitza left. He would also go on to win the Nobel Prize in 1951 for having been the first to verify Einstein's famous equation, $E = mc^2$! Seeing that Kapitza was publishing results similar to those obtained by Allen and Misener in his own laboratory, he asked them to publish quickly as well. I should add that the competition was necessarily tough: Jack Allen was one of the two young scientists hired by Cambridge to replace Kapitza when he was forced to leave. The other was Rudolf Peierls, a theoretician who would become famous for his work on quantum matter. Finally, since Jack Allen had worked in Toronto before coming to Cambridge, he brought along a student from there—Don Misener. To sum up, Allen and Misener, both Canadian immigrants at Cambridge, were working in part thanks to the material left by their predecessor and competitor, Kapitza, who was also an immigrant (from the USSR) at Cambridge, before being kidnapped by Stalin.

Let's come back to the date question. The Soviet physicist Andronikashvili claims in his book that Kapitza has priority over Allen because he submitted his article nineteen days earlier. But that's absurd: Allen hadn't simply reproduced Kapitza's work. His article with Misener contains a precise and detailed quantitative study of the flow of liquid helium through several tubes the thickness of a hair, which are called "capillaries." That's a considerable amount of work, requiring many more than nineteen days. Remember, each experiment began with producing the liquid helium necessary to cool everything else! They cited Kapitza out of intellectual honesty, and drew three astonishing conclusions from their measurements:

(1) Helium II flows much more quickly than helium I. A slight change in temperature is enough to alter the liquid's behavior completely.

(2) The rate of flow of helium II is essentially independent of the pressure at the ends of the capillaries; when you water your lawn, the output of your hose depends on the degree to which the faucet is open—however, helium II behaves very differently.

(3) This rate of flow also seems independent of the diameter of the capillary, which they had varied by a factor of a thousand. In a classical tube, whether or not it's for watering, the speed is proportional to the area of the cross section—it should have been 1,000 times greater in the biggest capillary than in the smallest!

Since the behavior of the helium II didn't correspond to any classical law, they simply stated that it was impossible to deduce a value for the viscosity. This conclusion was clear, rigorous, and accurate. They brought forth the first proof that helium II obeyed unknown laws, that they were dealing with a new kind of fluid. By contrast, the quadratic law invoked by Kapitza without precise figures would never be subsequently confirmed, and didn't prove that the flow was turbulent anyway. The scientific value of Allen and Misener's article is clearly superior to that of Kapitza's, so how is it that Kapitza got the Nobel Prize and not Jack Allen?

This injustice isn't only due to psychological or political reasons. There is an interesting underlying scientific problem: the theoreti-

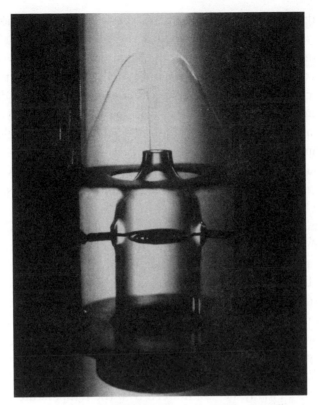

Figure 7. The fountain effect, photographed by J. F. Allen and J.M.G. Armitage in 1971. Courtesy of the School of Physics and Astronomy, University of St. Andrews, by the late Prof. J.F. Allen.

cal link between superfluidity and the Bose-Einstein condensation we spoke about in chapter 4. I therefore need to bring three more great men of science into the story: Fritz London, Laszlo Tisza, and Lev Landau. Herein lies another controversy!

Let's review the course of events. On February 5, 1938, that is, four weeks later, Jack Allen published a second article in *Nature*, this time with Harry Jones. He had just discovered that heating superfluid helium on one side of a capillary increased the local pressure, which raised the level of the helium and even made a fountain spring up. He named this new discovery the "fountain effect."

Given the setup of the experiment, the level of the liquid in the heated part should have dropped if the liquid had been classical. Observing the reverse prompted Allen and Jones into making obscure conjectures about the nature of the liquid. This time the theoreticians didn't have a choice: they needed a radically different interpretation for the behavior of this new state of matter.

One month was all Fritz London needed to send an article called "The lambda-phenomenon and the Bose-Einstein degeneracy" to *Nature*, and it was published on April 9. Born in Breslau (Wroclaw, in what is now Poland) in 1900, Fritz London was fleeing Nazi Germany, where he had worked with Erwin Schrödinger and founded quantum chemistry with Walter Heitler. Paul Langevin, Jean Perrin, and Edmond Bauer had welcomed him at the Institut Henri Poincaré in the heyday of the Popular Front, a leftist movement in France. London would stay in Paris until 1939, when he took up a professorship at Duke University in the United States. London had noticed that the temperature of the lambda point (2.2 K) was near that at which the kind of quantum condensation predicted by Einstein might have arisen (3.3 K). He had also understood that liquid helium was the site of high quantum agitation due to the low mass of its atoms. Finally, he had noticed that the temperature variation of its specific heat, the graph of which Keesom thought looked similar to the Greek letter lambda, resembled the one predicted for the phenomenon of quantum condensation by Einstein.

London had understood the behavior correctly, but at the time, few people believed in "Bose-Einstein condensation." Hadn't Einstein himself written: "The theory is pretty, but is there any truth in it at all?"[20] The theory of the state changes of matter was in its early stages. The Dutch physicist George Uhlenbeck, for example, didn't understand how two different liquid states could exist if it was impossible to observe their equilibrium. If you put ice in a glass of water and the temperature is 0° Celsius, the ice and the water are in equilibrium with one another. But such equilibrium is not possible between a classical gas and the same gas condensed in Einstein's sense. Progress on the problem had been made at the 1937 conference celebrating the centenary of Van der Waals, following a conversation between Einstein, Ehrenfest, and Kramers. London attended, to be sure, but the serious consideration of what still seemed to be a sort of pathology in the theory of quantum gases as the explanation for a new state of liquid matter was a real conceptual breakthrough. Landau, as we'll see, stubbornly refused to believe it, but first, the story has yet another twist.

Fritz London wasn't the only foreigner who fled to Paris. He had lured Laszlo Tisza, who had been born in Budapest in 1907, there as well. A former student of Max Born at Göttingen, Tisza worked with Teller under the direction of Heisenberg at Leipzig, but was arrested for thought-crimes by the Hungarian Nazi government. After fourteen months in prison, he sought refuge in a postdoctoral position at Kharkov—with Landau. In Paris, Paul Langevin brought him to the Collège de France with the help of the "French Committee for the Reception and Work Organization of Foreign Scientists." Working with Fritz London was easy, since getting from the Collège de France to the Institut Henri Poincaré was just a matter of crossing the Place du Panthéon. Jacqueline Hadamard was working at Paul Langevin's laboratory, and she helped Tisza stay in France until 1941, by which time the administration was able to get him an immigration visa for the United States. He was thus able to make it onto a boat, barely escaping antisemitic persecution, and take up the position that MIT had offered him.

Email really is a wonderful invention. In early 2001, after learning that Tisza was still active at age 94 as an emeritus professor, I looked online for his email address and found it easily. I then sent an email to him at MIT, on a lark, thinking it was like sending a message in a bottle. I wanted to get a better understanding of how this period of intense scientific activity unfolded in such an unstable political environment. Not only did he respond, but after a few emails, I was able to invite him to the École Normale Supérieure to give a talk on superfluidity. My laboratory is two hundred meters from the Institut Henri Poincaré, and we passed in front of it together. Tisza was walking slowly, and I was about to offer him my arm for support when we stopped for a moment in front of the large red-brick building on rue Pierre et Marie Curie. The memory of April 1938 came back to him. He and London liked to go for walks together, and it was during one of these walks that London had explained his idea: that superfluidity was due to the same kind of quantum condensation predicted by Einstein for gases, even though they were dealing with a liquid. Tisza, in turn, had a series of brilliant ideas which allowed for the explanation, on the basis of London's hypothesis, of all the known properties of superfluid helium, and even made some additional predictions possible. On May 21, 1938, again in *Nature*, he published a groundbreaking article describing what he called the "two fluids model." As its name indicates, this model maintains that liquid helium is a mixture of two fluids which are inseparable from one another but whose movements are independent, a revolutionary idea. How could one part of the liquid move without the other? Go mix water and alcohol, then try to move the water without the alcohol! But Tisza understood that this incredible possibility was a consequence of Bose-Einstein condensation. That seemed impossible even to London, but it turns out that Tisza was right.

This complicated story illustrates both the hesitations and controversies swirling around the agents of science, and the speed of events in a highly competitive field: all of these articles and even some other results generated by teams in Kharkov and Oxford, which I won't take the time to describe here, were published in the same volume of *Nature*, between January and May. Kapitza intervened soon afterwards to save Landau from Stalin's jails. Experimental research in England would soon be forced to come to a halt due to the War, and Landau would finally get back to work. In 1941, he published an article called "The theory of the superfluidity of helium II"[21] in the *Journal of Physics of the USSR*. In it, he developed an even more rigorous two-fluid model than Tisza's, which is still the basis for the understanding of superfluidity. The article has the great merit of introducing the concept of a "quasi-particle" to describe a quantum liquid. It explains both how superfluid helium flows, as well as its thermodynamic properties (how it heats, for example).

Even though fundamental to a large area of physics, the article is on the one hand flagrantly unjust to Jack Allen and Laszlo Tisza and, on the other, completely silent about Fritz London. It's worth a few more words about psychology, politics, and science.

This is how Landau's article begins: "Liquid helium is known for possessing peculiar properties, . . . among which the most important is the *superfluidity discovered by Kapitza.*" Certainly, Landau owed his life to Kapitza, but even Kapitza himself didn't lay claim to all the credit for the discovery. Landau continues by attributing the "suggestion" that there is a link between superfluidity and Bose-Einstein condensation to Tisza, rather than to its originator, London. Then, he writes: "The explanation advanced by Tisza not only has no foundation in his [i.e. Tisza's] suggestions but is in direct contradiction with them." Here he is being excessively severe with his former postdoc. Eight years later, after having corrected a slight error in his own theory, he clarified his opinion of Tisza:

"I'm happy to acknowledge my debt to Tisza, who . . . described helium II in 1938 by dividing it into two parts and attributing two velocities to them." Tisza's detailed article [*Journal du physique et du radium 1:* 165 and 350 (1940)] wasn't available at the time in the USSR because of the war, and I regret not having noticed his two earlier letters [*Comptes rendus de l'Académie des sciences* (Paris) 207: 1035 and 1186 (1938)]. However, his entire quantitative theory (microscopic as well as thermo- and hydro-dynamic) is, in my opinion, entirely incorrect."

So, we've learned two things. First, Landau still scorned his former postdoc's work; second, he hadn't read Tisza's articles which appeared immediately after Landau's short letter to *Nature.* On this last point, I wondered whether it was because Tisza had published them in French, which coincided with my own worries concerning international publication (see Chapter 12). But I had been told that, upon regaining consciousness after the serious traffic accident which had put him in a coma in 1962 and prevented him from receiving his Nobel Prize in person, Landau had spoken a few words— of French. To work through this confusing question, I contacted a former student of Landau's, Alexei Abrikosov, a famous physicist working on superconductivity who emigrated to the United States after the collapse of the Soviet regime. Abrikosov seemed shocked that I could have imagined that Landau, an absolute idol of physics in the USSR, might not have understood French. In his email of January 15, 2001, Abrikosov responded [sic.]:

"Dear Dr. Balibar,

Landau was very able to languages. He knew German, English, French, and Danish. Therefore he could read Tisza's papers in French, the more so, that Lifshitz, whom he often ordered to read papers, instead of doing that himself, didn't know French. What concerns your second question, I doubt it, since I was present in the hospital practically all the time of his regaining consciousness (the first word he pronounced was my name) but I haven't heard a single word in French (my French is far from good but sufficient for distinguishing it from English).

Sincerely yours,

Alex Abrikosov

It's touching to see a student so filled with admiration for a teacher who had nonetheless tried to destroy his own theory of the penetration of a magnetic field in a superconductor, even though he was right, and would even receive the Nobel Prize for it in 2003. But Abrikosov had given me a plausible explanation of why Landau was perhaps not aware of Tisza's early work on helium II: Landau used the services of Lifshitz to read scientific literature. E. M. Lifshitz would be the coauthor with Landau of a series of physics books that physicists throughout the world still use today. Had Lifshitz forgotten to relay Tisza's detailed studies to his mentor Landau? Regardless, the influence of Landau would become enormous: if Landau said it, it had to be true; superfluidity had been discovered by Kapitza alone, and his theory owed nothing to Tisza.

Not to Tisza, and not to London either. But Landau knew all about London's work. He had met London a few years before. Fritz London also had a brother, Heinz, with whom he published. Landau cited that brother when he got to the mathematical formulae that described the fountain effect: "The formulae 6.1. and 6.4 were deduced already by H. London (Proceedings of the Royal Society, 1939), starting from Tisza's ideas."

So Landau deliberately ignored Fritz London.

How was that possible?

Landau held London's mention of Bose-Einstein condensation against him. Throughout his own work on the subject, Landau systematically avoids making reference to it. For a long time I thought that Landau refused to consider the idea that a property predicted by Einstein for a perfect gas could apply to a liquid. In a perfect gas, by definition, the atoms are completely independent from one another and don't interact among themselves. By contrast, in a dense liquid, the atoms are very close to one another and thus constantly interact. Besides, the first extension of the idea of Bose condensation to a gas (with weak interactions among the atoms), wasn't completed until 1947 (by another Russian, Bogoliubov), and then its generalization to a liquid would take Penrose and Onsager another ten years. In fact, as Tony Leggett (who got the Nobel Prize in 2003 at the same time as Abrikosov) says, it's not clear whether it's possible to go smoothly from the Bose-Einstein condensation of a liquid to that of a gas. (Bose-Einstein condensation in a gas was discovered and proven with certainty in 1995.) The link between superfluidity and Bose-Einstein condensation is still not, in fact, entirely uncontroversial; my colleague Yves Pomeau, for example, still considers them to be two different phenomena which aren't necessarily related.[22]

I thought, then, that at the root of the injustice done to London by Landau, there must have been a scientifically motivated reason, even though Landau was notorious for his lack of scruples when making his views about physics public. I was surprised to learn from Lev Pitaevskii that the reason was indeed scientific, but slightly different from the one I had first come up with. Lev Pitaevskii was Landau's 25th student; Landau recruited his students on the basis of a tough exam and then wrote their names down in a large notebook. So, Evgueny Lifshitz was number 2, Tisza number 5, Abrikosov number 12, Pitaevskii number 25, Sasha Andreev, with whom I exchanged polemics on helium crystals for fifteen years, number 32.

Like Lifshitz, Lev Pitaevskii played a large role in writing the series of books that dramatically affected 20th century physics. He also wrote a very interesting article in which he discusses the circumstances surrounding the publication of Landau's article in 1941.[23]

When the Soviet Union collapsed, Abrikosov emigrated to Argonne in the United States, but Pitaevskii preferred to go to Trento, in Italy. So during a conference organized in Trento, I had another chance to clear up my mystery; I asked Pitaevskii why, in his opinion, Landau was so persistent in denying any link between superfluidity and Bose-Einstein condensation, and therefore refused to cite London. Pitaevskii helped me think back to the time when scientists didn't really even understand superconductivity itself. According to Pitaevskii, Landau was convinced that superfluidity and superconductivity were essentially the same phenomenon. Why? How? It's a mystery, but Einstein's reasoning applied to the particles called "bosons" (because their statistical properties are determined by Bose's laws), while electrons belong to the opposite group, "fermions" (determined by Fermi-Dirac statistics). According to Landau, since helium and superconductors exhibited the same property, despite being constituted by particles which obey different statistical laws, neither should be tied to questions of quantum statistics. So, Landau was too early! He didn't have the theory of superconductivity, which was established on the basis of London's work. In 1957, Bardeen, Cooper, and Schrieffer, all Nobel Prize winners, of course, figured out that by associating in pairs, electrons become bosons and thus undergo Bose-Einstein condensation.

This also helps us understand why the polemic between London's partisans and Landau's was focused on the question of whether helium's light isotope, helium 3, was superfluid or not. In effect, the helium 3 atom contains one neutron less than helium 4, and so is a fermion. If London was right, liquid helium 3 should not be superfluid, at least at comparable temperatures. That, indeed, is what D. W. Osborne, B. Weinstock, and B. Abraham discovered at Argonne National Laboratories in the United States in 1949, once they had high enough quantities of helium 3 to perform the experiment (helium 3 is a by-product of the study of tritium, which was undertaken with the goal of producing the H-bomb). London and Tisza didn't hide their glee. Later, in 1973, David Lee, Douglas Osheroff, and Bob Richardson discovered that atoms of helium 3 could also pair up and obey Bose's statistical laws, like the electrons in superconductors, thus allowing helium 3 to become superfluid, but at temperatures a thousand times lower. As for Landau, he would also turn out to be right, but about other aspects of superfluidity, for example, the explanation of the maximum speed at which a superfluid can flow, as well as the description of the thermodynamics of a superfluid at non-zero temperatures in terms of "quasi-particles."

In 1954, London died prematurely of a heart attack. A prize was founded in his memory, and in 1960, the jury responsible for finding a winner chose . . . Landau. I was a bit surprised when I discovered that.[24] The two men were radically opposed to one another, both in their ideas and their interpersonal behavior. But thinking about it now, I believe that Landau nevertheless had a certain respect for his adversary. He simply didn't want to cite a work whose central hypothesis he didn't accept. If he hadn't died prematurely, London might have shared the Nobel Prize with Landau, especially since Einstein himself had suggested London for the supreme award. London probably deserved a Nobel for chemistry as well. In any case, as for Jack Allen, I think he was the victim of a flagrant injustice, for which I see only one explanation: Landau's prestige was such that if he said it, it had to be true; superfluidity was therefore discovered by the person who invented the word for it—Kapitza, and him alone. As for Tisza, at the end of a talk given on the occasion of the centenary of the Hungarian Society for Physics in 1991, he wrote: "If history has a lesson, it is that the 'winner takes all' attitude deprives one of the pleasure of being the heir to the best of different traditions, even while avoiding their intolerance against one another." In sum, London and Allen suffered at the hands of a general tendency to remember only a single victor in the great historical battles of science. Tisza found himself stuck between two conceptions of superfluidity which took decades to unify.

This is how science moves forward—with conceptual breaks, the invention of words, multiple authors, and occasional polemics. But

89

where has this history of hard work led us in the end, in terms of the applications of liquid helium? Among other things, liquid helium is used today to cool every medical imaging setup using nuclear magnetic resonance (the MRI scanners in hospitals), and CERN is building a giant 27 km ring of superfluid helium in which to submerge the magnets of its new particle accelerator. Perhaps it will play a key role in helping physicists revise the standard model of particle physics, and we'll soon learn of new words invented to describe mysterious particles, yet to be discovered.

Crystals and Glasses

"I expected something more impressive!" That was my father's reaction when, as a young director of research (a position equivalent to associate professor), I brought him to my laboratory.

It was in the evening, and I had taken care to open the door without turning on the neon lights on the ceiling right away. Sitting on its optical table, the laser pierced through my cryostat with its red beam; orange displays and green dials shone from the control panel; a camera transmitted an image onto a TV-screen of what I considered to be the best crystals in the world, or at least the coldest and purest—hexagonal helium crystals. My pumps were humming, and my notebooks were spread out in front of this multicolored Christmas tree, evidence of all my hard work. Even the top of my cryostat was giving off cold vapor. But I suppose reality didn't measure up to the expectations of a father with an excess of love for his son.

In November of 1986, I took one of the photos that *Physics Today*, the monthly journal of the American Institute of Physics, was going to use on its cover and wrote on the back, "This photo is one possible symbol of ten years of my life." That photo sat enthroned on his fireplace mantle for another ten years. But why did I spend such a long time slaving over crystals?

I've always liked simple questions. In this case, the question was,

Figure 8. Transfer of liquid helium, rue Lhomond, 2004. Photo by Th. Bouet.

Figure 9. Reprinted with permission from *Physics Today* (Vol. 40), © 1987, American Institute of Physics.

Why do crystals have facets? And if that took me ten years of work to answer, despite the help of several especially brilliant students, it's no doubt because the question was a difficult one.

Look at salt with a magnifying glass; you'll see that the little crystals are cubic. Candied sugar also makes regular polyhedral shapes,

93

bounded by small symmetrical facets. As for quartz crystals, look at them closely in a mineral store, or better, go visit the amazing mineralogy collection at the University of Paris-6 at Jussieu! Quartz crystals are polyhedral spikes whose tip and prismatic body are hexagonal. Remarkably, these symmetries—cubic for salt, hexagonal for quartz—reveal the internal structure of the crystals: their molecules are arranged in such a way as to present the same symmetries as the form of the whole, as was explained by the abbot Haüy, then later Auguste Bravais, and finally Pierre Curie over the course of the nineteenth century.[25] This correspondence of symmetry is so precise that quartz crystals, whose silica molecules are arranged with a symmetry which isn't precisely hexagonal, themselves have a shape which also isn't, since the facets are arranged in two groups of three.

These facets appear so frequently on crystalline surfaces that their existence is often taken to be the definition of the crystalline state itself. And yet the forms of some crystals—I mean the natural, raw, unpolished forms in which they grow—are perfectly rounded in all directions, while the atoms which constitute them are indeed arranged regularly at the nodes of a regular network, which is the periodic repetition of a basic pattern (and that's the correct definition). That's what we discovered in 1980: our helium crystals were rounded when warm, but the more we cooled them, the more their surfaces changed from rounded to plane—into so-called "facets." And we wanted to know why.

Incidentally, is all solid matter crystalline? No. Like a liquid, a solid is an example of matter which is dense, not diluted like a gas. But solids and liquids have different mechanical properties. In effect, a solid resists an application of shear stress: if I push the top to the right while I pull the bottom to the left, a solid will become a little deformed, but if I release the stress, it returns elastically to its original form. There are hard solids like, say, a piece of iron at room temperature, and soft solids like—it will surprise you, but I'll clarify what I mean immediately—a jelly or a soap foam.

Blow on a collection of soap bubbles and you will see that it is elastic; shake some jelly lightly, and you'll see the same thing. On the other hand, a liquid flows if you subject it to shear stress: place some oil between two plates, then move one plate relative to the other. The oil flows and won't bring the plates back into their original position when you let them go. But be careful! If you put too much stress on a solid—and it doesn't take much for a lather or a jelly—it becomes, as we say, "plastic": that is, it too is irreversibly deformed, and doesn't return to its initial shape when the stress is released. So, *the solid or liquid character of a material is not an intrinsic property* but depends rather on the force applied. Moreover, we know that it's easy to make many things flow besides suds and jellies. A metal can be made to flow without melting: that is how we make iron wires by forcing the metal through a hole (we "extrude" it).

Let's come back to our question: are all solids crystalline? By mentioning foams and jellies, I've already given the answer, which is negative. Whatever they're made of, suds and jellies are not regular, periodic arrangements. It's best to place glass among these non-crystalline solids, which deserves some explanation. If I twist a pane of glass, it resists like a solid and eventually breaks, but it doesn't flow like a liquid. However, a glass is not a crystal; it is, as scientists call it, a "jammed" liquid, whose atoms or molecules are distributed in space in a disordered way, blocking their motion as in a traffic jam. It resembles gels such as jellies or yogurts, only harder.

Now, both the word "glass" and the word "crystal" have two different senses, one common and the other scientific, so it's easy to get lost. A glass is both an object for drinking and a material whose inherent disorder interests me.

That reminds me of a phone call I got one day from a journalist for *Astrapi*, a magazine for children. She had questions about different sounds in daily life, and an association called Science Contact had suggested that she call me.

"Why do crystal glasses sound different from ordinary glasses?" she asked me.

These kinds of questions can be delicate, and I was happy to find a plausible answer quickly. It's a matter of vibration, the pitch of which is higher if the material is more rigid. When we make a toast, the impact of one glass against another causes a resonating oscillation in the shape of the glass: the top edge, which is circular to begin with, becomes slightly elongated, then flattened, then elongated again, and so on. With a high-speed camera, we can record the breakdown of the movement. If the glass has a stem, and we hold it by the stem so that the vibration of the top is not impeded, the sound is pretty, since the vibration isn't dampened, and it lasts a while. The material of "crystal" glasses is more rigid, and so the pitch of the emitted sound is higher, just as a violin string produces a higher sound if you stretch it more, making it stiffer.

"But are crystal glasses made of glass or crystal?"

Glasses which we call "crystal" are indeed made of glass, that is, a disordered solid material. "Crystal" in the ordinary sense is simply a glass loaded with lead oxide, which makes it denser as well as stiffer; sound propagates more quickly through it than through ordinary glass, and so it produces a higher sound. To me, as a physicist, the expression "crystal glass" is familiar, but a contradiction in terms.[26] Happily, that doesn't stop me from liking to pour a great Burgundy into one, a Vosne Romanée that smells of violets, for example. I still hold my glass by the stem, to hear the length of the sound, which goes along with the length of the taste.[27]

The physics of glasses, in the scientific sense, is one of the most active areas of my discipline—as is the area of disordered systems more generally. It's worth a long detour. We'll come back to crystals afterwards. For simplicity's sake, let's say that a glass is a liquid that has cooled down so quickly that it didn't have the time to crystallize. We make window glass by quenching molten silica, that is, silicon oxide in the liquid state. If we were to give it time, it would have crystallized as quartz. But by pouring it over a bath of quickly flowing cold water, we end up with the glass with which we are familiar.

Why doesn't the silica crystallize? For crystallization, the molecules need the time to order themselves in space, and that time can be long, for two reasons which often go together. On the one hand, the liquid might be highly viscous, while on the other, the periodic configuration might not be noticeably more stable than other states. In this case, the glass behaves as if it were hesitating about whether or not to take on a periodic configuration, which doesn't clearly seem to optimize its energy. This situation is especially interesting, and we find it when matter is the site of contradictory forces. Physicists, who like metaphors, speak of "frustration," evoking the conflicts of everyday life.

Let's imagine a group of people of two different colors; half are red and half are blue. Suppose they want to reorganize themselves, but each person prefers to be next to someone of the other color; it could be a group of men and women who prefer the opposite sex, if you like. Now suppose we ask these people to distribute themselves by fours around square tables: they will easily find a solution that satisfies everyone, with two blues and two reds per table, alternating. We can generalize to a large square network, in which we don't want any two boxes of the same color next to each other, and anyone who has seen a chessboard will know that there is a simple solution to the neighbor problem. But try coloring a network of hexagons—a honeycomb—in blue and red and you'll see that it's impossible to avoid having hexagons of one color with neighbors of the same color. So, to return to the human metaphor, we're going to make some people unhappy; there will be satisfied people and dissatisfied ones, that is, a certain level of general frustration. The group of people will have multiple possible configurations, none of which will be truly satisfying. This situation, in which contradictory tendencies lead to a multiplicity of possible configurations, and so to great difficulty in finding one configuration which is better than many others, is characteristic of glasses.

Moreover, this situation has a very interesting consequence: glasses age—they have a history, unlike ordered crystals, which have none. When you cool a liquid whose constituents have contradictory tendencies, it gets frozen as a glass, usually ending up in a state which is not the one with the lowest energy, and thus not in a state of equilibrium. Since its temperature still isn't zero when it does this, some thermal agitation remains, and so it continues to evolve, albeit slowly, as it seeks out the equilibrium state which would optimize its energy. Because of its internal contradictions, glass doesn't stop evolving. Conversely, crystals are stable configurations of matter that have no reason to evolve towards a more stable state—they have already achieved it. I implied in passing that glass undergoes internal motion, even though this motion is very slow. I thus suggested that glass is a solid if you look at it over a short period of time, but it is a liquid if you observe it for a long time. The difference between a solid and a liquid, therefore, depends not only on the force being applied but also on the duration of observation. Complicated, certainly, but surprising, and therefore interesting.

Some physicists take the frustration metaphor even further, citing Tolstoy: "Happy people have no history"; in other words, frustrated glasses do. That's pretty well put, and I myself am tempted to extend this business about motion induced by contradiction to other domains. Thus, when the painter François Rouan speaks of "disgusting beauty," he is using a shocking phrase to illustrate his feeling that in painting, contradiction—or "collision" as he says— can generate an emotional response, adding a temporal dimension to the work. In a painting, if we look at one element of its construction we see something different than when we look at an opposing element. As we keep looking, the way we see evolves, but we might have the impression that it's the painting that changes. In fact, if I take more pleasure in research than teaching, it's also because when the researcher butts up against the unknown, his ideas are forced into motion, whereas the teacher has a tendency to limit himself to what seems established, coherent, and therefore fixed (though I'm aware that the teacher has good pedagogical reasons for adopting that attitude).

At long last, let's get back to my crystals. Unlike glasses, they are ordered periodic systems in space. In our laboratory, we've shown that their polyhedral facetted shape is a consequence of that order, and that changes in temperature are accompanied by changes in shape only because shape depends on the agitation of matter. When our crystals were warm enough, their shape was completely rounded, like a drop of liquid.

The shape I'm talking about is still the raw shape in which crystals grow, the shape of salt crystals produced when we let salty water evaporate from a plate, or that of minerals we find in nature, not

Figure 10. The colder they get, the more facetted helium crystals become.

what we can get by polishing (like the shapes of precious stones transformed into jewelry). So the shape of a crystal is made up of facets regularly placed on its surface, and the colder the temperature, the more the crystal is facetted, that is, the more its surface is covered with small flat regions oriented in increasingly varied directions.

The four photographs here show helium crystals at lower and lower temperatures: from top to bottom, at 1.4 Kelvin, then 1 Kelvin, 0.4 Kelvin, and 0.1 Kelvin. Remember that the Kelvin scale is the absolute scale for temperature, and that this scale is simply displaced by 273.15 degrees from the Celsius scale (0 K = −273.15 °C). One Kelvin is therefore very cold, but helium atoms exert only a very slight attraction on one another; the energy of their interaction is very weak. As a consequence, even around 1 K the properties of helium still depend on temperature; it's like a very light object being weakly attracted by the Earth—the slightest breath is enough to move it. The topmost photograph, therefore, shows a completely rounded crystal; you would think it was a drop of liquid. And yet we know that it's a helium crystal, a hexagonal arrangement of atoms; when we cooled it, we observed its shape flatten out on top, then become a hexagonal prism with six lateral facets around an upper hexagonal facet, and finally six more facets appeared, tilted at about 60° and cutting across the lines of the earlier prism.

We were not the ones to discover solid helium; that was Willem Keesom, at Leiden, in Holland, in 1926. However, when he searched for the crystallization point of liquid helium, he measured thermodynamic quantities, but he couldn't see anything: "[. . .] There was nothing peculiar to be seen," he wrote. It's true that helium crystals are especially transparent, and with the naked eye, it's just as difficult to distinguish their surfaces as when looking at an ice cube submerged in pure water. When Étienne Rolley, Claude Guthmann, and I managed to take these pictures in color, we therefore felt we had done a bit better than Keesom, but since it was sixty years later, that was to be expected. In fact, we were astonished that our method of capturing images was so good. It consisted in lighting our crystals with light dispersed by a prism—a sort of rainbow—and then fiddling with a small black mask at the focal point of the lens, which produced the image in our camera. Each pair of facets constituted a small helium prism which showed up as a different color. But we hadn't thought of this method ourselves; it was suggested to us by an engineer, Leonard Tanner, who was friends with Pierre-Gilles de Gennes, and who was taking advantage of retirement to photograph rotating drops of water in his garage in Bristol. I think that the success of our work has as much to do with these images as with its rigorous scientific content.

We did discover, however, that helium crystals have a variable number of facets, but we weren't the only ones. We were in competition with at least two other teams: Konstantin Keshishev and Alexandr Parshin in Moscow, and Judd Landau and Steve Lipson in Haifa. Each team had its own experimental methods and its own theories. The period from 1978 to 1994 was a lively one, in part because each conference gave us the chance to see who was progressing faster than the others, but also because we could see, on the one hand, whether we were observing the same phenomena and, on the other, whether we were interpreting them in the same way. By 2004, the competition between myself and my Russian colleagues had become collaboration, thanks to the Finns, with whom we found ourselves gathered all together around more powerful machines than our own. I sealed my friendly understanding with Alexandr Parshin by coauthoring a lengthy journal article with him.[28]

Certainly, one of the hardest questions on which to make any headway was whether the appearance of facets is linked solely with the decrease in thermal agitation, or whether helium's quantum properties also play a role. That subject is too vast for this book, but it's not hard to get an idea of why thermal agitation has consequences for the shapes of crystals.

When you look at a crystalline facet with a tunneling microscope—the little marvel invented by Gerd Binnig and Heinrich Rohrer in Zurich, and which won them the Nobel Prize in 1986—you see that the surface is flat, with atoms nicely arranged next to one another. The crystal is a regular stack of layers of atoms, of which the exposed facet is the last layer. I suppose everyone has at some time or another seen a fruit display. Vendors usually arrange their oranges in regular piles, at least those which are round enough to allow for it; and indeed, atoms of the same kinds are all exactly the same size. On a larger scale, piles of oranges are crystals, and they terminate with a final, flat layer. In the morning, before the first customer has arrived, the last layer of oranges is full, the vendor hasn't removed any fruit from it to sell, and the surface of the pile is flat, even if it also may be tilted at an angle for display.

But why does the temperature change anything about the order of the surface? Matter is agitated by fluctuations, of which temperature is one measure. The warmer a body is, the greater its internal agitation. As long as the temperature of a crystal is low, the last layer of atoms is flat; but if the agitation increases, the same thing happens to the crystal that happens to the pile of oranges if you shake it: some oranges will jump on the surface of the pile from one place to another, creating holes and bumps in the beautiful arrangement we started with. The surface of the crystal becomes disordered when the temperature increases; it's no longer a plane in the crystalline network, and the facet disappears from the shape. Physicists call this disordered state of a surface "rough," evoking the sea in high winds. I like the metaphorical intrusion of the sea into the physics of crystalline states.

Why focus on helium, then, to study all of that? For the pleasure of working near absolute zero, or of observing crystals through multiple windows, each joint of which is prone to leakage? Of carting around countless heavy containers of nitrogen and liquid helium through the years? Reaching and maintaining temperatures of less than one Kelvin; combining optics, electronics, and cryogenics; making measurements carefully without adding heat to the system, in a cave that was insulated from vibrating too much due to trucks passing overhead, all that wasn't easy, and took us several years. I picked helium not for love of the challenges it posed but because it provides ideal situations for studying the general properties of matter.

Since helium is the smallest of the noble gases, its atoms attract each other less strongly than the others. This near-absence of interactions means that if we want helium atoms to arrange themselves in liquid form, let alone as a solid, rather than remaining diluted in a gaseous state, the helium atoms must be drastically cooled, so that nearly all their thermal agitation stops. To purify liquid helium, then, all that's needed is to cold-filter it: all other matter is frozen at such temperatures, and thus easily separable. The purest, most inert, and simplest of the elements: these are the main characteristics of helium which make it a model system. There are facets on all crystals, but in helium, we can better control the parameters on which the facets depend. In addition, the last several decades have seen the development of low-temperature physics, and measurement techniques—thermometers and pressure sensors, for example—have been developed which are especially sensitive in the neighborhood of absolute zero.

After all that work, we were able to generalize what we learned from helium to the rest of crystalline matter. So, for example, we were able to compare helium and silicon—*the* crystal medium of modern electronics. We found similar behavior for surfaces with intriguing properties, which are tilted and look like atomic staircases. And speaking of staircases, when Pawel Pieranski discovered at Orsay that some liquid crystals—not the ones used for watch displays but other, cubic ones, made of an incredible network of soapy membranes—displayed up to 60 different types of facets at room temperature, we revisited the ideas we developed using helium with Philippe Nozières. After a few months of discussion, we figured out why these liquid crystals display facets grouped into "devil's staircases," a metaphor meaning that this facetted staircase has an infinite number of steps, which get narrower as they go up, and which it would be rather difficult to climb.

So I think that helium plays a special role in physics. Besides, we wouldn't have been able to begin the study of the internal cohesion of water if we hadn't first perfected an acoustic method of studying the behavior of bubbles and cavities in liquid helium.

Helium is, however, absent from a scientific book which has long moved me, Primo Levi's *The Periodic Table*. The book begins in this way:

"Argon. There are the so-called inert gases in the air we breathe. They bear curious Greek names of erudite derivation which mean 'the New,' 'the Hidden,' 'the Inactive,' and 'the Alien.' They are indeed so inert, so satisfied with their condition, that they do not interfere in any chemical reaction, do not combine with any other element, and for precisely this reason went undetected for centuries. As late as 1962 a diligent chemist after long and ingenious efforts succeeded in forcing the Alien (xenon) to combine fleetingly with extremely avid and lively fluorine, and the feat seemed so extraordinary that he was given a Nobel Prize. They are also called the noble gases—and here there's room for discussion as to whether all noble gases are really inert and all inert gases are noble. And, finally, they are also called rare gases, even though one of them, argon (the Inactive), is present in the air in the considerable proportion of 1 percent, that is, twenty or thirty times more abundant than carbon dioxide, without which there would not be a trace of life on this planet. The little that I know about my ancestors presents many similarities to these gases."[29]

Levi describes his family, which anti-Semitism kept apart from the rest of the population of Piedmont, before describing his life as a chemist, without a word about his time at Auschwitz, a horror he describes elsewhere. Thus, Levi cited all the noble gases except for the simplest of them, doubtless because its etymology is less poetic: helium gets its name from the Sun (*helios*), where it was observed for a long time before we learned that it was present on Earth (where it is constantly produced by the disintegration of uranium). Nevertheless, for me it's a little element which guides us, like the diminutive Petit Poucet (or Tom Thumb) in the tales by Perrault, through the jungle of the properties of matter.

8

God, Helium, and Universality

〜〜〜〜〜〜〜〜〜〜〜〜〜〜〜〜〜〜〜〜〜〜〜〜〜〜〜〜〜〜〜〜〜〜〜

I remember a slightly awkward conversation I had once with Luc Ferry in 1999. At the time he was still president of the *Conseil national des Programmes*, at the Ministry of Education. I had just been asked to represent the physics community and learned that Ferry had asked our council to compose a "single manual" (or textbook) of knowledge. The idea was nothing less than to present, in a unified manner, the whole of knowledge in a form accessible to students in middle and high schools. And, the council continued pointedly, it should "make sense." We were supposed to be able to present in condensed and organized form the "basics of knowledge," which would, of course, delight both students and their teachers. And it was supposed to be 200 pages long. Of course! The task of compressing all of knowledge into a 200-page manual should be easy!

Somewhat dumbstruck (the work had already begun, and they were waiting for my contribution), I launched into a long rant, explaining first that I didn't see how we were going to unify the whole of knowledge when physics itself wasn't unified. In front of eight philosophers, I dared to contest the very idea that the whole of knowledge was capable of unification. Such an assumption seemed to me, in effect, to be based on a belief to which I refused to adhere: could this unifying principle of all knowledge be anything other than God Himself?

The great undertaking fell flat. However, I would have rather liked to discuss calmly this problem of *unity*, which had made me think of the physical concept of *universality*.

I'm certainly not the only one to notice that we've only unified three of the four fundamental forces of nature in a global theory: the "standard model" of interactions. In the nineteenth century, Maxwell and others unified electricity and magnetism. Scientists in the twentieth century did succeed in unifying electromagnetism, or more precisely quantum electrodynamics, that is, the quantum theory of the interaction between charged matter and light, with the strong interaction responsible for the cohesion of nuclei and the weak interaction that plays a role in certain nuclear disintegrations. But this comprehensive theory does not include the gravitational interaction, the force which makes apples fall and is the most familiar to us. The unification of the whole set of forces was a challenge for the twentieth century and remains one for the twenty-first.

Now, some of my theoretician colleagues claim that if we managed to unify the theory of the four fundamental interactions of matter, we would have constructed a "theory of everything." This assertion is something I've always considered to be a debatable kind of intellectual imperialism; indeed, it suggests that understanding nature can be reduced to identifying the elementary particles and determining their interactions.

Obviously, I don't think that's true.

Understanding how water freezes, how a piece of iron changes shape under stress, or how a turbulent atmosphere fluctuates are three examples of questions I challenge the champions of the so-called theory of everything to solve in terms of unified fundamental interactions. No: a group of particles has properties which are not understandable simply on the basis of the properties of those individual particles or even pairs of such particles. A group has properties that its individuals do not. Understanding the properties of single atoms isn't enough for understanding those of apples.

Pascal was subtler when he said: "Since all things . . . maintain each other by a natural bond, . . . I hold it to be impossible to understand the parts without understanding the whole, or to understand the whole without especially understanding the parts." I don't think there is *one* theory of complexity, despite what you may sometimes read on the topic. As for some phenomena, like turbulence, which are difficult to understand (see the next chapter), I don't think that recognizing the complexity of the phenomenon will help establish a theory of it, nor that understanding the nature of the elementary particles and of their interactions will help us to calculate the drag on a boat or what triggers a storm.

So, physics is far from being unified.

Nonetheless, it's not a collection of heterogeneous pieces of knowledge. First, it gets its power from a rigorous scientific method common to all its practitioners, and which is generally considered to have been invented by Galileo, even though he had precursors in the Arab world. We observe, we develop a theoretical model, we confront the model with new observations, we adjust the model if necessary, and so on. The triad of observation, modeling, and verification of predictions is the framework of our entire science.

Above all, though, as Jacques Treiner says,[30] the procedure common to all physics consists of "making something simple by searching for what is identical in the diverse, or, as in Jean Perrin's words, *by searching for the invisible simple behind the complicated visible. And it finds it.*" The search for unifying generalizations, for a single conceptual framework to describe diverse phenomena, is a fertile attitude of mind. From Maxwell to Einstein, the history of physics proves this abundantly. Indeed, I've studied crystals myself for so long because I was seeking to identify universal laws governing state changes in nature.

In a letter to Marcel Grossmann dated from 1901, Einstein wrote: "It is a wonderful feeling to grasp the unity of a collection of phenomena which appear to immediate sense perception like completely separate things." He was twenty-two years old. So was it for pleasure that he worked so much to unify physics? His discovery of the quantization of light was indeed the result of such an approach. In 1905, he published a series of five theoretical papers that would have a great impact on the century. In one of them, he introduced "quanta of light," which are also called "photons" (*Annalen der Physik* 17: 132 [1905]). There, Einstein pointed out the difference in treatment which physicists reserved for waves and matter: "According to Maxwell's theory, energy is to be considered as a continuous spatial function for all purely electromagnetic phenomena, hence also for light, while according to the current conceptions of physicists the energy of a ponderable body is to be described as a sum extending over the atoms and electrons. The energy of a ponderable body cannot be broken up into arbitrarily many, arbitrarily small parts, while according to Maxwell's theory . . . the energy of a light ray emitted from a point source of light spreads continuously over a steadily increasing volume." Einstein applied to light waves a calculation usually reserved for a gas of particles and thus derived a number. This number was the energy of the radiation divided by its frequency and by Planck's constant. He then interpreted this relationship as the number of quanta—the number of grains of light, if you like. In his conclusion, finally, he proposed that the photoelectric effect (the emission of electrons by a metal affected by incident radiation of light) be studied to verify the existence of these quanta of light. The test was conclusive and the existence of photons definitively accepted. Thus, by seeking to unify the continuity of waves and the discontinuity of matter, Einstein developed a more general theoretical framework than his contemporaries and allowed for the explanation of new phenomena such as the photoelectric effect (and many more besides).

Was Einstein's motivation purely aesthetic? It's possible, but I doubt it. It seems to me that in physics generalizations are required by friction—the conflict between rival theories. According to some observations, for example, light seemed to be a wave, but under other circumstances to be a collection of particles. Short of admitting that light was not a single and unique physical phenomenon, a theoretical generalization capturing both aspects at the same time was required. It's therefore the friction between the wave theory and its rival corpuscular theory that led to Einstein's theory. Without this friction, why would one need to generalize? Speaking of which, some people believe that Einstein quantized light *in order* to explain the photoelectric effect, whereas he quantized light and *then* discovered that this phenomenon allowed for understanding the photoelectric effect.

Now is a good time for a digression about quantum evaporation. Adrian Wyatt, whose postdoc I was in 1977, had the good idea of naming the phenomenon thus, but I was the one who discovered the phenomenon itself several years earlier during my doctoral work. I had noticed that a small burst of heat propagating through helium was prone to push out atoms upon arriving at the open surface, to evaporate them, in a way. Then I discovered that at low enough temperatures these atoms were coming out with a remarkable minimum speed, 79 m/s.[31] And that's what Phil Anderson[32] had predicted in 1969, in the interesting case where heat in helium is decomposed into "rotons," the particles Landau had invented to explain the thermodynamics of superfluid helium. In the photoelectric effect, a quantum of light (a photon) that runs into a metallic surface ejects an electron whose kinetic energy is the difference between the energy of that photon and the binding energy of the electron with the metal. In my case, the rotons ejected the atoms with a kinetic energy which was the same simple difference between the energy of the incident particle and the binding energy; the figure of 79 m/s was derived in a straightforward way. So my burst of heat was a well-defined packet of particles (Landau says "quasi-particles") with a well-defined energy. It was no doubt less surprising for physicists, seventy years after Einstein, but all the same, at twenty-eight, I was rather proud to have shown so directly that heat, too, could be quantized.

Let's come back to the unity of the sciences. To my knowledge there is no friction between physics and biology, and so no need for a more general theory that captures both physics and biology at the same time. Of course, that doesn't prevent the two sciences from complementing each other to describe, say, the micromechanics of biological molecules or the functioning of the molecular motors thanks to which our muscles contract or our cells divide. Nor does it prevent physicists and biologists from teasing each other in friendly exchanges.

That several attempts to unify the sciences have turned out to be fruitful proves only that science sometimes makes progress by resolving some of the difficulties it confronts in this way. But that's not always the case.

Nor should I be misunderstood as to the end result of the successive generalizations that mark the progress of physics. They illustrate the way in which scientific truth evolves in our discipline. When a theory is accepted at one time, it is rarely declared false later on; in other words, it's rarely invalidated. Rather, it is usually generalized in a form which captures the earlier state of the theory and simply allows us to go farther in our understanding of nature—that is, in the interpretation of additional phenomena which have, in the meantime, enriched our description of nature. Maxwell's theory, in fact, was itself a remarkable generalization, even more often cited as one than Einstein's General Theory of Relativity when epistemologists describe the progress of science. It is worth a detour as well.

Thirty years before Einstein, the Scottish scientist James Clerk Maxwell was attempting to unify electricity and magnetism. To do this, he wrote a group of four equations which permit us to calculate, for example, what magnetic field is induced by an electric current in a coiled wire or, inversely, what electric current is induced in the coil if we move a magnet in its vicinity. Researchers today will appreciate the fact that he published his equations in 1873, two years after becoming director of the famous Cavendish Laboratory of Cambridge: at the time, the job of administering a large laboratory didn't preclude active engagement in research!

In fact, though, the most remarkable part is that Maxwell's equations contained more than he thought he put into them. Among the solutions of the equations were electromagnetic waves, the remarkable and unanticipated possibility that an electric field and a perpendicular magnetic field propagate by oscillating in any medium, including in a vacuum. Maxwell's equations predicted the propagation of light and of radio waves, which Heinrich Hertz discovered soon afterwards, in 1887. But the arrival of Maxwell's theory and *a fortiori* of Einstein's quantum theory happily don't prevent us from using Descartes's laws of optics, which do not require Maxwell's theory. This so-called geometric (or ray) optics has its own domain of application, and retains it despite those developments.

Later, in the same laboratory, Dirac established an equation that described both the relativistic and the quantum characters of the electron, leading to the discovery of antimatter. But fortunately, that didn't prevent anyone from using Schrödinger's more elementary equation to describe the evolution of quantum particles when their speed is low compared to the speed of light.

The level of my own research has never brought me to the height of those giants, but I have firsthand knowledge of my own desires and pleasures, and those I felt when I encountered the concept of universality in physics. In the end, it's what has interested me the most in my study of the shapes of crystals, and I'll say a few words about it. It will also be a good opportunity to revisit the distinction between continuity and discontinuity, but from a very different point of view than that of Einstein's quantization of light.

This universality concerns certain state changes of matter which are called *continuous* as opposed to the changes which are more familiar to us, such as crystallization or boiling of a liquid, which are *discontinuous*. When liquid water changes to vapor, for example, its density jumps from a large value for the liquid to a small one for the gas without passing through a series of intermediate densities, and so the state change is discontinuous. These discontinuities also pose interesting problems, but first let's address the *continuous* shape-changing of my crystals.

I and my closest colleagues and students in this long adventure—Étienne Wolf, François Gallet, David Edwards, Étienne Rolley, Claude Guthmann, and François Graner—weren't content to just take pretty pictures for scientific journals. We made measurements and compared them to the predictions of a mathematical theory elaborated by Philippe Nozières as we proceeded. Philippe's calculations were complex, and I was happy that he went at them fearlessly. For his part, he was glad that we were taking care of the cryogenics necessary for our experiments, since that, too, was a lot of work. In order to compare the model to reality, we measured the curvature of

111

the crystals at the location and temperature at which the facets appeared, and also how passing from a rounded shape to a facetted shape changed the growth rate of our crystals, and how the radius of a round facet shrunk to zero as the temperature approached that of the "roughening transition" where the facet disappeared. And we measured a number of other quantities which are difficult to explain without equations. As for the theory, it made use of some of the most recent refinements in the physics of state changes of matter, having to do with "renormalization"—which earned the American Ken Wilson the Nobel Prize in 1982. This theory claimed that the shape changes of our crystals occurred in a way similar to that of many other state changes in nature. We had discovered that the mathematical model describing crystalline surfaces led to the same equations as the model describing several other systems, which are called "two-dimensional," since they are thin layers, such that two spatial coordinates describe their position.

I'm sensitive to the aesthetics of a theory that is capable of reaching something general after starting from specific cases. I still find it rather beautiful that these two-dimensional systems for which we predicted the same behaviors were as diverse as a thin magnetic layer becoming magnetized when it's cooled, a collection of positive and negative electric charges becoming conductive, a liquid film which crystallizes or another which, instead, becomes superfluid, and so on. Thus, for example, the curvature of our crystals was supposed to vary with temperature in the same way as the magnetization of magnetic layers, according to the same square root law. That prediction seemed to us to be so astonishing as to justify all the years of work, and we did indeed verify it. One of the great successes of statistical physics in the 20th century was to classify all the continuous state changes in nature as a function of a few simple criteria. What, for example, is the dimension of the space in which the change takes place? A three-dimensional volume? A two-dimensional layer? What kind of order arises? Thus, even though these systems appear to be completely different, we demonstrated and verified that they are governed by universal laws. We thereby contributed to the unification of a small corner of physics.

Let's get back to the crystallization and boiling of water, which we took up after studying the shape of helium crystals. When I make water boil, then, its density jumps from a large value to a small one; that's the typical example of a *discontinuous* state change. In the same way, if I cool water and it freezes, it goes discontinuously from a liquid state (where the positions of the molecules are disordered) to a crystalline state (in which the molecules are arranged in an orderly way), and where the symmetries of the spatial arrangements are different from the symmetries of liquid water. Water doesn't go through any intermediate state between liquid disorder and crystalline order.

How does a physical system jump from one state to another? We all learned in primary school that 0 degrees Celsius (0°C) is the "temperature of melting ice." In effect, if we start cooling water, its temperature falls, but once ice cubes start appearing, it stabilizes at zero as long as any liquid water remains. An equilibrium is established between the two states of the water, and as long as the two states coexist, the temperature is stable: 0 °C. But does the first ice cube appear once the temperature reaches zero? Not at all! If we perform the experiment in a very clean container with very pure water, we can easily go a few degrees below zero. A certain amount of "supercooling" is required before the first bit of ice will form. Water in clouds is in the form of droplets in suspension, not touching any wall, which allows it to stay liquid down to about −20 °C, though this temperature depends on the level of pollution. Below 0 °C, water is in a state called "supercooled," which is stable only as long as it isn't disturbed too much (we therefore call it "metastable"). Some people say that the water in mountain lakes is pure enough to stay liquid rather far below zero as well. It is in a fragile equilibrium, like a pencil standing upright, which can be knocked over with the slightest breath. The whole lake crystallizes immediately if it is disturbed ever so slightly—maybe not by dipping one's toe in, but certainly by throwing in an ice cube.

The liquid and crystalline states of water are therefore like two sections of a box separated by a barrier. To move from one section to the other, that is, from one physical state to the other, the water

113

must jump the barrier. The equilibrium temperature (0 °C) corresponds to the situation in which the two sections are at the same height. Below zero, things proceed as if the box is tipped: since one section is lower, the water will be more stable there. If I tip the box more (read: if I cool the water much further below zero), a moment will arrive when the water in the higher section will go over the barrier into the lower section, especially if I shake it a little (though in fact, thermal fluctuations agitate the water even without my intervention). At what temperature does that occur for water? The supercooling of water has interested physicists for a long time, but it seems as though it's impossible to prevent water from turning to ice below −40 °C. Is this very cold temperature the point at which the barrier no longer resists the water passing from one section to the other? It's possible, but not certain. This kind of property depends on the structure of liquid water, which is of a well-known complexity (ice is simpler). And it happens that at the beginning of the twenty-first century, while there are some physical theories of impressive sophistication, we don't know the structure of water well enough to be certain. That's what attracts us to this area of research.

If our knowledge of the structure of water is insufficient to predict how far we can prevent it from freezing, we should expect the same difficulty in understanding boiling. Taking a lot of precautions, we can prevent water from boiling up to around +200°C, much hotter than the official boiling point, 100°C. A related phenomenon is that of *cavitation*. Temperature and pressure are the two physical quantities on which the state changes of pure matter depend. If instead of heating water at ordinary pressure (one atmosphere, around one bar), we lower the pressure at room temperature, bubbles will also appear, and the liquid water will pass to the gaseous state. This phenomenon is similar to that of boiling but has a different name: we call it "cavitation," and it is very important, for example, in hydraulics. Whirlpools form behind the propeller of a boat, in the centers of which the pressure drops. It's like the eye of a tornado. In fact, when the atmospheric pressure drops in your region, and the weather service tells you about it, it's simply because you are in the middle of a giant atmospheric whirlpool passing through.

All that is related to what is called "Bernoulli's law," which describes the conservation of energy in hydrodynamics. It was established by Daniel Bernoulli, a Swiss physicist[33] who had been invited to honor Saint Petersburg with his presence by Catherine I.[34] So, bubbles form at the centers of whirlpools created by the propeller. They lower its efficiency and, by imploding against its side, they can make small holes in it, leading to progressive deterioration. It's this destructive capacity of bubbles that doctors exploit when they use ultrasound machines to destroy kidney stones. Finally, some researchers in plant biology think that the reason trees cannot grow above a certain height is that the vertical sap column pulls with all its weight on the sap at the top in such a way that if the weight is too much, it produces cavitation, which destroys the cells. Apparently you can even hear the sound.

Cavitation, then, is an important phenomenon. In ordinary water, cavitation happens easily. Think back to the trees. In the sap column, the pressure decreases with the height. When you dive, the same phenomenon takes place but in reverse: the deeper you go, the more you feel the pressure increase. It does so by 1 bar every 10 meters, and you've perhaps guessed why: all the water above is weighing down on you. Alternatively, imagine that you were able to move up inside the channels carrying the sap; in that case, the water is in fact pulling, and the pressure is 1 bar (atmospheric pressure) at ground level, and so 0 bar at a height of 10 meters, −1 bar at 20 meters, −2 bars at 30 meters, and so on. It may sound surprising to speak of negative pressure, but it's no different from a positive stress—it's a stretching, if you like. Now, it's well known that if you pull on a piece of wood, it will break if the stress is large enough. The same goes for a liquid. But it's also known that if wood already has some small fractures, it's more fragile. In the same way, ordinary water, which contains lots of defects—especially microbubbles of air—breaks, that is, cavitates at slightly negative pressure: −1 bar for seawater behind propellers, and up to −10 bars in the tallest trees, apparently. In fact, the flow of the sap plays a role as well, especially in a drought, when there is a large amount of evaporation through the leaves. Trees are therefore able to resist cavitation down to much greater negative pressures.

However, in a very pure liquid, there is neither air nor any other dissolved gas, so cavitation takes place only if the stress applied really manages to separate the molecules, and for that, much greater negative pressures are required. In very pure water, which we are studying at the moment, the cavitation limit is predicted to be −1,500 bars, an extremely powerful stress which we know how to produce with exceptionally strong ultrasound (which beats looking for a tree 15 km high!). But in fact, we don't really know the limit for the rupturing of pure water, since we don't adequately understand the structure of water or the forces which attract the molecules to each other and give it its internal cohesion. It's pretty irritating not to know the structure of the liquid that is indispensable for life. It's that irritation that launched Frédéric Caupin and me into this new adventure.

So, the end of the twentieth century saw the solution of the problem of *continuous* state changes, but we don't fully understand some *discontinuous* state changes of matter—and those of water are at the top of the list. If we could add a piece to that puzzle, it would allow us to engage more easily with our biologist colleagues when they tell us that, while physics was certainly the queen of sciences in the twentieth century, it is their turn to take the throne.

Cyclists and Butterflies

In general, the wind blows from the west in the Chevreuse valley, and so we are heading into it when we leave, every Sunday morning, for a bicycle ride towards the Rambouillet forest. It's better that way, since then the return is with the wind at our backs and seems less hard. Wind is the enemy of the cyclist: on a plain, we use up practically all of our energy fighting wind resistance.

I'm trying to save some energy before the next rise, where I'll feel gravity pulling on my 80 kilos, and where I know Paul will take off again, and Michael, who's light, will easily leave me behind. So I try to wedge myself in behind the wheel of a large person with decent surface area: my son Lucien for example, or Rino, or even Rémy, but not Sophie, not Marc, nor Benjamin. They, in turn, prefer to protect themselves behind me, at least if I'm not trailing behind. But now, I need to feel the direction of the wind, the speed of which combines with that of the cyclist. There is a precise position in which I can pedal effortlessly, in which I almost feel sucked along; that sheltered position is only right behind the cyclist in front of me if the wind is blowing exactly head on, or if our speed is substantially greater than the wind speed. In the former case, our little peloton stretches out single file. In general, though, the wind comes from a bit off to the side, and to be properly sheltered I need to move my front wheel up next to the back wheel of the person in front of me, fanning out. But with a difference of five centimeters in one direction or the other, the wind hits me and I get as tired as if I were in front. It takes thousands of kilometers of training to easily find the position of rest in the aerodynamic trough behind another cyclist, but at 40 km/h, an amateur like me has to be careful, since even accidentally brushing my front tire against the back tire of the cyclist in front of me could prove fatal. Everyone does what he can to save energy without too much risk, and for once, I find it useful to have some knowledge of fluid mechanics.

I've worn a helmet ever since a butcher from Antony, running a stop sign in his big Mercedes, fractured my face. My helmet has straps which run alongside my ears and clasp below my chin, and these straps whistle like the stays on a sailboat in the wind. The whistling reminds me of the screeches that accompanied my attempts to play the flute, when I was around thirty and wasn't able to direct my breath properly over the bevel of my Yamaha's mouthpiece. I've since gone back to the piano, but I've also learned that the whistling of the wind comes from the instability of the streams of air going around an obstacle, whether it's a stay, a strap, or a bevel. They don't flow with a constant trajectory but rather oscillate from one side to the other, creating alternating vortices. There are beautiful photographs that show these vortices, which are pushed along by the air current, widening and forming the undulating sides of a "street," which was analyzed by Theodor von Karman.[35] The frequency with which these alternating vortices are created increases with the average speed of the flow around the obstacle: the whistling is higher when the air moves faster. That is how the wind sings: it modulates is speed over obstacles. As for me, since I encounter less wind when I'm properly sheltered behind Lucien, I have a trick to help me find the right position: I listen to the von Karman "streets" in the straps of my helmet. A little too far to the left, and the whistling is high. Too far to the right, same thing. At just the right spot, the sound is low. Of course, I also listen to my thighs, which burn if it's too difficult, and the rhythm of my heart, which panics when I'm not wedged into the right spot for the hellish pace being set. None of those indicators tells me immediately, and the wind fluctuates, the road turns, and so I constantly have to readjust the position. The pros can do it by instinct; I take the opportunity to do some physics.

When the wind blows across us and we take turns relieving each other this way, I think about wild ducks, who, with more room in the sky and no oncoming cars, arrange themselves in a V. Apparently, despite their lack of education in fluid mechanics, those great birds also feel the air turbulence in their teammates' wakes. They relieve each other the way we do on our bikes, and go much faster as a group than individually. Fish in schools also practice applied hydrodynamics, to protect themselves from predators. However, as for ants, whatever their motivation, I think they do chemistry in order to follow in the paths of their siblings and neighbors.

After a century in which research on quantum mechanics and relativity have almost monopolized the covers of scientific journals, the study of turbulence is coming back onto the scene. In effect, scientists are realizing that they don't understand it, or at least not enough. What's more, turbulent flows are the rule rather than the exception: in the atmosphere of a star, in that of the Earth, in the oceans, behind a car or a boat, inside a motor, or in a chemical reactor, the flows are all turbulent. So thinking about turbulence provides an opportunity to distinguish *what we know that we don't yet know* (how to calculate, for example, the drag force that limits the maximum speed of a boat or a car) from *what we know that we'll never know* (how to predict the weather more than two weeks ahead).

A flow that isn't turbulent is called "laminar": it is stable in time and space. We can visualize streams of air or water flowing around an object by injecting dyes, and if the flow is slow enough or of small enough dimension, or if the fluid is very viscous, these streams have a regular stable form. Let some liquid honey or glycerin run, and you will generally get a laminar flow. If you don't mind throwing it out afterwards, you can verify this by injecting ink into your honey and running a spoon through it. By contrast, run a finger through soapy water or a spoon in your coffee cup, and you will see vortices forming on either side of your finger or spoon—the flow is unstable, turbulent.

Figure 11. Behind a round obstacle, a flowing fluid creates vortices alternating from one side to the other. This phenomenon accounts for the whistling of the wind in the stays of the masts on boats. © ONERA.

There are, nonetheless, various degrees of turbulence. At relatively low speed, the flow can oscillate periodically; that's what happens with the wind in the stays, which whistles at a rather well-defined frequency and thus produces a sound which we can more or less reproduce by whistling or singing. But at very high speeds, or around a very large object, the wind doesn't make any sound of an identifiable pitch: the flow is so turbulent that it becomes chaotic and only makes noise.

A bit of musical acoustics should help us understand the concept of chaos, which is very important in contemporary physics in general, not just the physics of turbulence. A pure sound is a simple periodic oscillation of air pressure, which can be emitted by various cheap electronic gadgets, for instance. A musical sound is always periodic, but composed of the superposition of a fundamental oscillation, which determines the pitch we perceive, and its harmonics, that is, oscillations at frequencies that are double the fundamental one, triple, quadruple, and so on. This is all according to proportions which determine the timbre of the instrument; thus we can easily distinguish an oboe from a flute, since the sound of the former is much richer in harmonics than the latter.

A chord, of course, is what you get when you play several notes at the same time, and is therefore the superposition of several sounds; however, it is always the sum of a *finite* number of oscillations at identifiable frequencies. It's not easy to sing chords, but some singers manage it—in Sardinia for example. I also know a member of the French Academy of Sciences, who is also a remarkable musician, and who can sing and whistle at the same time; so, he can play two-part Bach fugues without a musical instrument. But the important thing, and what interests us here, is that a noise is the superposition of an *infinite* number of oscillations at different frequencies, and physicists say that its spectrum is "continuous." Unlike a sound which, being periodic, is always reproduced in the same way and is therefore perfectly predictable, a noise is called "chaotic," that is, unpredictable. Even if you record a noise up to a certain point, you will be unable to predict what follows, since it is random. A strongly turbulent flow is chaotic: as unpredictable in time as it is in space.

One question that saw a lot of progress at the end of the last century is that of the different ways in which a stable flow at low speeds can begin to oscillate, and then become chaotic at high speeds. There are several "routes to chaos," and I saw up close how one of them was discovered.

Around 1980, when I was starting to wonder why crystals had facets, Albert Libchaber was starting to heat cubes of liquid from below. Albert had been my thesis supervisor, and I owe him a double debt: he was the one who really encouraged me to develop my curiosity, and he was also the one who pushed me to define my own subjects for research. He worked in the room next door with Jean Maurer, an especially original engineer, who had built a small plastic box, fitted with a small tube used to fill it with liquid helium. We all appreciated the charms of that model liquid for answering general questions. Maurer's little box was also furnished with a heating wire underneath and a heat sink on top, as well as a few microthermometers. At low heating power, the heat was simply transmitted by thermal conduction from the bottom to the top and the fluid remained immobile. At higher power, the fluid started to move; this "convection" movement increased the heat conduction between the top and the bottom, and its source was as simple as Archimedes' principle. In general, heating a liquid in effect produces an expansion, and in the presence of the Earth's gravity, the hot liquid, which has become lighter, rises, cools when it comes into contact with the upper wall of the container, becomes denser, and goes back down. Combined with buoyancy, thermal expansion destabilizes the fluid to produce a convection roll. The shape of the roll depends on that of the box: if it is roughly cubic, only one roll is formed, and it can rotate in either direction. If the box is rectangular, roughly twice as wide as it is high, a double roll arises, in which the fluid rises in the middle and goes back down on the sides. Finally, if the box is much wider than high, multiple long rolls are formed and may be aligned parallel to one another or organized in a network of hexagonal cells. You can observe such roll structures by heating a layer of oil in a pot. I've also already mentioned (chapter 3) that this thermal convection is at work in the Earth's mantle, dragging the plates of the Earth's crust and triggering earthquakes.

Libchaber and Maurer observed the same kind of convection, a rather well-known phenomenon in fact, in their plastic cube. But by increasing the heating power even more, they observed that the temperature began to oscillate in time. It oscillated because the roll-shaped motion of the liquid was itself oscillating. At first they were surprised to see that as they kept increasing the heat, the oscillation period doubled, then doubled again, then again and again until finally the periodic motion of the rolls became chaotic fluctuations. Doubling the period meant dividing the frequency in half, and if Libchaber and Maurer had been able to listen to their box, they would have heard the sound get deeper and deeper, an octave at a time, before becoming a sort of formless noise. Later, acoustics researchers reproduced this phenomenon and effectively heard the "cascade of period doublings," as its inventor, Mitchell Feigenbaum, named it. As for Albert, the experimental discovery of one of the "routes to chaos" allowed him to realize one of his dreams, a career in America. This kind of oscillating convection is a widely prevalent phenomenon, which has since been discovered in several other areas: electronics, optics, chemistry.

You've no doubt noticed that I used the plural "routes"! That is because there are indeed others. In 1971, or ten years before the experiments performed by my neighbors down the corridor, David Ruelle and Floris Takens had shown that for a dynamic system to become chaotic, it need only be able to oscillate in three different modes which are coupled to each other. David Ruelle wrote a remarkable book on the subject (*Hasard et chaos* [Odile Jacob, 1991]; *Chance and Chaos* [Princeton University Press, 1993]), but my own way to understand it is to come back to the planets of my Provençal nights. If the Earth were the only planet revolving around the Sun, the Earth-Sun pair would form an oscillator with a stable period (one year), the future of which would be perfectly predictable. No doubt you think this is not only true but also independent of the presence of other planets. It certainly is on the scale of human life, and even on the scale of our civilization. But, in fact, each planet rotates with its own period, and all of their movements are paired up, since gravitation is at work between all of them. One planet's passing near another slightly modifies the trajectories of both, and these modifications are amplified with time. According to Ruelle and Takens—and they're right—it only takes three planets paired in this way in order for the movement in general to be chaotic. As Jacques Laskar explains, on the scale of a hundred million years, the movements of the planets of the solar system are chaotic and unpredictable.

So, over the course of the last three decades of the twentieth century, scientists have shown that the great Lev Landau was wrong when he claimed that the appearance of chaos in the turbulence of a fluid was due to the progressive outbreak of a larger and larger number of oscillations, all at different frequencies. Three is enough, or even a single one which divides itself. Even the greatest physicists are sometimes wrong.

But let's return to the notion of prediction, so we can look at the memory of a physical system. The movement of a clock is so regular that if you know its state at any one moment, you can predict its state at any later time. A clock has a perfect memory of its past: its future state is entirely determined by its prior state. On the other hand, a chaotic system loses all memory after a certain amount of time, which is why it's so difficult to predict the weather. That difficulty and loss of memory come from what is sometimes called the "butterfly effect," which deserves some clarification.

The butterfly effect gets its name from a talk given by the meteorologist Edward Lorenz in Washington in 1972: "Predictability: Does the flap of a butterfly's wings in Brazil set off a tornado in Texas?" There he emphasized that the flap of a butterfly's wings in some location was more or less enough to change the subsequent evolution of the whole atmosphere completely, so great is its sensitivity to initial conditions.

We can understand the sensitivity of a chaotic system to initial conditions by imagining a billiard table with a round block at the center. We roll a ball towards the block, and it bounces off to one side or the other depending on whether it hits on the left or right of the block. It then bounces off the side cushions and traces a certain trajectory along the mat of the table. If we were to try to obtain the same trajectory by rolling the ball a second time, we would need to be very, very careful, since the slightest difference in direction on the initial impact will lead to a difference in direction for the first bounce, which will be enough for the ball to hit a different side cushion after a few more bounces, leading to a radical difference in the rest of the following movement. A billiard table with a block in the middle is therefore chaotic: we can't get the same trajectory for two throws of the ball which hits the block unless the precision of the throw is infinite. Various configurations of these chaotic billiard tables have been studied by Yakov Sinai, a mathematician working today at Princeton: some don't have a central block but simply a stadium-shaped edge, with a straight part connecting two circular parts.

We can also make chaotic number series. Let's start with a number between 0 and 1; say, 0.33. Then double it: 0.66; and double it again: 1.32. But whenever the result goes over 1, we knock it back down between 0 and 1 by removing the integer part, which in this case yields 0.32. And we keep going: 0.64, then 0.28, then 0.56, then 0.12, et cetera. Now, let's begin again with a neighboring number, for example 0.34. Instead of getting the series

0.66, 0.32, 0.64, 0.28, 0.56, 0.12, 0.24, and 0.48,

we get:

0.68, 0.36, 0.72, 0.44, 0.88, 0.76, 0.52, and 0.04.

We see that these two sequences of numbers diverge rather quickly and become completely different. That comes from the combination of stretching and folding to which we subject these numbers, by doubling them and then reducing them to below 0 and 1. In the same way, on my chaotic billiard table, the block pushes the trajectories apart while the side cushions knock them back inside. In a turbulent flow, the lines of current are also stretched and folded by the vortices.

Mathematicians recognized the importance of this "stretching-folding" combination by giving it a metaphorical name: "the Baker's Transformation."

Have you ever made layered puff pastry? I myself follow the "quick" method, from the bakery Lenôtre: I spread out a layer of pastry dough in a shape three times as long as it is wide, which I dot with butter two-thirds of the way along. To do that, I recommend flattening the butter between two sheets of plastic wrap with a rolling pin, then putting that "sandwich" into the refrigerator for five minutes so that you can unstick the butter from the plastic. Then, I fold the dough in thirds, rotate the resulting square by 90°, roll it out again, then fold again, roll again, et cetera. After repeating that operation a few times, each particle of butter is far away from its initial neighbor; the butter has been chaotically dispersed between the various layers.

In sum, what arithmetic and pastry have in common is that they show that chaotic behavior can result from simple, perfectly deterministic laws, which are neither complicated nor probabilistic as we might have expected. Getting back to meteorology, the reason we can barely predict the weather more than a week in advance is that the movements of the atmosphere and the oceans are chaotic. It should be noted that the fundamental Navier Stokes law governing hydrodynamics is, nevertheless, perfectly deterministic. So are the laws of gravitation, which are at work in the movements of the planets.

What moves the atmosphere? It's the heat of the sun: the air is hotter at ground level than at high altitude and hotter at the equator than at the poles, and as in Maurer's experiment, that triggers gigantic convection movements. What's more, since the Earth turns, the vortices have a favored direction, which is why the dominant winds come from the west in France. It's more or less the same for the oceans: the water is hotter and saltier at the equator than at the poles, which sets off gigantic ocean currents. The Earth's rotation plays a role there as well, which is why the celebrated Gulf Stream rotates towards the west in the north Atlantic, while a current of extremely cold water goes back down to the east, along the American coast. The Earth is a giant thermal machine which transforms energy from solar radiation mainly into winds and sea currents, though also into rain and snow, with well-known consequences for vegetation.

Until the middle of the twentieth century, we were content to observe and describe the weather in the various regions of the globe or, on a longer timescale, the dominant climate. For predicting the weather, we were reduced to taking the average from previous years. How is it that we do it better nowadays and predict for at least the next few days? There are three main reasons. First, we have satellites which allow us to observe the whole planet: even though they could still be improved, of course, we always know the wind, the pressure, the temperature, the humidity, the presence and type of clouds. Second, we have computers powerful enough to handle the collection of these data, which can quickly calculate how the ocean and atmosphere are evolving from the conditions measured at an initial instant. To do these calculations, the computer uses the laws of hydrodynamics, which, I repeat, are well known. Finally, we have a number of models that describe the instabilities that can appear here and there: models of the formation of clouds and the way they affect precipitation, the local temperature, the reflection of solar radiation, and so on. The good meteorological models are constantly improving, though no doubt everyone has his own opinion about the trustworthiness of their predictions. My personal impression is that predictions up to twenty-four hours in advance are relatively reliable, except during some parts of the year or in certain regions where the weather is especially unstable. But we all know the rather modest reliability of weather forecasts several days ahead. No institute makes a serious attempt to predict the weather more than a week in advance—and the origin of that uncertainty is the sensitivity of chaotic systems to initial conditions. We can dream of satellites which will one day measure atmospheric conditions every meter instead of every kilometer, or even better, but the evolution of the terrestrial system will still be chaotic. We might manage to predict the weather with greater reliability a few more days in advance, but never more than two weeks. Even if a forest of hawk-eyed satellites were constantly registering the positions of all the butterflies which ever so slightly disturb the local air by fluttering from one flower to another, even if a fleet of computers could handle the corresponding mass of data in less than twenty-four hours, we still wouldn't be able to predict the weather more than two weeks ahead. It's not the butterfly that unleashes a storm on the other side of the globe, as some wrongly twist Lorenz's thought into saying, since the engine that creates storms is still the Sun; it's rather the presence of a butterfly in one place rather than another which can suffice to make a storm happen farther away, earlier on, later, or somewhere else entirely, in an unpredictable manner.

Nonetheless, we shouldn't think that meteorologists are useless—far from it. For one thing, I repeat, their short-term predictions are pretty reliable and are always improving; for another, the atmosphere and the oceans are the scene of phenomena linked to larger structures, which have been identified and on the basis of which meteorologists make predictions. The best-known example is El Niño, the great mass of hot water that sometimes crosses the Pacific from west to east and arrives to warm the coast of South America, leading to a proliferation of shrimp and other life. Even though that instability is random, we can predict it six months in advance. Moreover, when climatologists average out the fluctuations over longer periods and look at the evolution of the climate, they free themselves from unpredictable fluctuations and their predictions become deterministic again. That fact is, thankfully, reaching the front pages of newspapers, since the general situation is worrisome: climate models predict a global warming, the cause of which is human activity. (We are emitting more and more carbon dioxide.) Scientists are only hesitating at the moment between predicting a rise in 3° or 6° Celsius, not because of uncertainty about the evolution of the ocean and atmosphere but mainly because we don't really know how the population will evolve nor the way it will continue to consume energy and produce "greenhouse gases."

While I was thinking about all of these scientific problems, I heard Daniel Cohen declare on the radio that we shouldn't expect economists to make predictions any more than we should expect meteorologists to do so! This great economist—a professor at the École Normale Supérieure, a member of the Prime Minister's council of economic analysis, and the winner of various international prizes—seemed to me to be downplaying his responsibilities just a bit. Just because economists' predictions are notoriously uncertain, should they be excused from making them? Isn't it a bit much for him to compare economists to meteorologists, who do in fact make predictions, taking into account all the precautions we've just discussed? I felt I gained a better understanding of what Cohen meant when he gave a seminar in our laboratory in January of 2005.

The economic state of the world depends on phenomena which amplify its fluctuations and are responsible for chaos in a way that is indeed comparable to atmospheric phenomena. For example, economists are interested in the evolution of income per inhabitant. No doubt from a desire to simplify, they first consider that in a market economy, each individual acts in pursuit of his economic interest, that is, the money he or she can earn given his or her means and abilities. If everyone everywhere had the same ability to choose, globalization would imply a homogenization of incomes on the planetary scale. But that isn't the case. If you are ten times as rich as an African peasant who hires out his labor force on a field of cotton, you can invest and earn much more than ten times what he earns.

The difference between your incomes has every likelihood of growing irreversibly. When George Soros plays the financial markets in billion-dollar amounts, his expected profit is not commensurate with that of a small investor; he is capable of influencing currencies all by himself. The economy is therefore governed by laws which are called "nonlinear," since the effects are not proportional to the causes. And nonlinearity is an essential ingredient in chaos. If the laws of hydrodynamics weren't nonlinear, we could superimpose several motions upon one another without them influencing each other.

I'm not sure if we know exactly how to predict whether a system is chaotic or not, but when there are many parameters, and it is governed by nonlinear laws, it's highly likely that it is. And that's not only true of the economic state of the world as a whole; but if you look at the fluctuations in the stock exchange or in the prices of commodities, it simply jumps out at you that you're looking at random noise, regardless of the general tendencies or the cycles that can be superimposed on it. These fluctuations in financial markets have something else in common with hydrodynamic turbulence: they both feature what are called "intermittent" phenomena. Chance in economics isn't as simple as it is in playing dice. We know very well that there are serious crises, and that stock exchanges crash. But like storms and earthquakes, these large-scale events are much less rare than they would be if they were simply random phenomena. Serious crises like these perturb markets for some time, just like earthquakes and serious storms have long-term after-effects.

Thus, the evolution of economic variables is probably chaotic, like that of the atmosphere, that is, it is sensitive to initial conditions and therefore unpredictable. But regarding the economy, Gérard Jorland is right to point out that the initial conditions are very difficult to measure, if only because of fluctuations of currency values.[36] Moreover, we at least know the laws of hydrodynamics and gravitation, whereas as far as I know, we don't have a reliable global mathematical model of the economy that would allow us to calculate its evolution starting from a given instant. So I don't think that economists can really calculate the evolution of their variables.

That doesn't mean economics can't have a certain scientific aspect, at least with respect to the rigor with which economists criticize their own theories.

"But if you don't make predictions how do you respond to the prime minister when he asks your opinion?" I asked Daniel Cohen. He responded that economists give advice on the basis of empirical knowledge of the effects that can be produced by such and such a political decision in similar situations. So I understood that economists are sort of in the same situation that meteorologists were in before 1950, when they could do no better than make predictions on the basis of averages from past years. No doubt they have no other choice. They still need to know whether the situations they consider to be similar are similar enough.

Another question therefore arises: if economists had a mathematical model of the economy, could they make reliable predictions? Would it be possible at least in the short term, as with meteorology? Everything would depend on the degree of chaos in the economy.

The atmosphere is not the only turbulent fluid, far from it. I mentioned the surprising case of the planets of the solar system, whose chaos is manifested on the scale of hundreds of millions of years. We might also consider terrestrial magnetism. In effect, we all tend to believe that compasses have been indicating North in the same direction forever and will continue to do so. But terrestrial magnetism comes from the convection movement of the liquid iron core which fills the inside of the Earth at depths between 2,900 and 5,150 km. Like any electric current, the movement of molten metal generates a magnetic field. In 2007, this effect was reproduced in a laboratory by Stephan Fauve and his French colleagues, who confirmed that the turbulence of the Earth's core is essential to the spontaneous appearance of its magnetic field.

But this internal turbulence of the Earth has a remarkable consequence, which has been verified. Our magnetic field is not at all stable on the scale of geologic time. It has reversed itself in a random manner around twenty times in the last few million years, as if the flow of the enormous mass of liquid iron suddenly changed direction. But rest assured that on the scale of a human lifetime, your compass probably won't change direction and send you to Antarctica when you're trying to get to Greenland.

Depending on the chaotic system under consideration, then, the timescales according to which the system's behavior is unpredictable can vary from a week (the atmosphere) to a million years (the Earth's core) to a 100 million years (the entire solar system).

If we recognize, therefore, that the state of the economy, whether national or global, is chaotic, it's understood, naturally, that it is unpredictable, but on what timescale? This is a question which seems very interesting to me, and I doubt that economists are able to answer it. But who knows? Perhaps great progress in this domain is on the way. If so, the comparison between economics and meteorology can be made more precise.

And so, as chaos regularly makes the front pages of scientific journals at the beginning of the millennium, the noise of political storms is heard on the radio, and the daily papers are filled with stories and commentaries about populations stretching beyond and folding back behind fragile borders. A singular coincidence.

Other Apples

As much as I try to convince myself that the Bible is a poetic text, that what is written there isn't meant to be taken literally, I really have a hard time reading Genesis, at the beginning of the Old Testament, without reacting strongly. The text is sacred for both Jews and Christians:

> And the LORD God commanded the man, saying, Of every tree of the garden thou mayest freely eat: But of the tree of the knowledge of good and evil, thou shalt not eat of it: for in the day that thou eatest thereof thou shalt surely die. [. . .] And when the woman saw that the tree was good for food, and that it was pleasant to the eyes, and a tree to be desired to make one wise, she took of the fruit thereof, and did eat [. . .]
>
> And they heard the voice of the LORD God walking in the garden [. . .]
>
> And the LORD God said unto the woman, What is this that thou hast done? [. . .]
>
> Unto the woman he said, I will greatly multiply thy sorrow and thy conception; in sorrow thou shalt bring forth children; [. . .]And unto Adam he said, Because thou hast hearkened unto the voice of thy wife, and hast eaten of the tree, of which I commanded thee, saying,

Thou shalt not eat of it: cursed is the ground for thy sake; [. . .] In the sweat of thy face shalt thou eat bread, till thou return unto the ground; for out of it wast thou taken. . .[37]

Basically, Eve was curious. She wanted to know good and evil in order *to become wise*. As for Adam, this authoritarian god blamed him for listening to Eve, born from his rib to be "an help meet for him". . .

But yet, isn't curiosity a magnificent quality?

Many interpret this text in a restricted sense, equating the good with chastity and the bad with sexuality, whence the idea that by sinking her teeth into the apple, that is, by being curious about the desire she inspired in Adam, Eve committed the original sin. Even allowing such a simplistic interpretation, I confess that I rebel against such an austere and unrealistic morality. What is more legitimate, more respectful in a loving relationship, than to be interested in the other's pleasure? Should we develop our sensibilities and share our happiness, or rather condemn ourselves to suffer and so make those around us suffer as well?

But the "knowledge of good and evil" in this text is a way of speaking about knowledge generally, of course. Eve stakes out a claim for the autonomy of thought—the right to understand *herself* what good and evil are. And anyway, the text speaks of wisdom— the understanding of everything. Defying the taboo, Eve expresses the selfish desire to think and therefore understand: good and evil to start with, maybe, and then the rest. But Yahweh reserves that right for himself alone.

Eve should have been happy following the divine law, but she wanted to do her own research, which displeased Yahweh!

If that's what sin is, then I'm all for it.

Understanding how the world was formed, how life could have appeared in a corner of it, and how nature behaves are the objects of my healthy curiosity, and these are the apples that this book invites you to sink your teeth into. And I invite you to share my taste for the apples of knowledge all the more since contemporary science is still finding fresh, new and exciting ones.

Speaking of apples and the way some researchers today confront

Figure 12

the mysteries of how forms appear in nature, I especially like the work of Stéphane Douady and Yves Couder.

Douady and Couder work in the same physics laboratory as I do, at the other end of the corridor. At the beginning of the 1990s, in this place where scientific inquiry doesn't get in the way of either fantasy or aesthetic pleasure, they started looking at pinecones.[38]

Wrapped around the base are scales that seem to be arranged in spirals. Our eyes see these spirals by associating scales with their neighbors, but we can do it in two different ways, depending on whether the spirals rotate to the left or to the right.

Let's count the spirals, as Douady and Couder did. In figure 12, you will see 8 in one direction and 13 in the other. If you were to count the spirals of a cedar, instead, whose cones are smaller and whose scales are closer together, you would see 5 in one direction and 8 in the other. As for the cypress, they are in the 3/5 family. Strange numbers, aren't they?

As did Douady and Couder, start over with a sunflower, a nice

Figure 13

one, open and mature, not damaged by any accidents during growth. You should see 21 spirals in one direction and 34 in the other, or 34 and 55, or 55 and 89, or even 89 and 144 if the flower is really large; but rarely any different numbers. Definitely strange! Did sunflowers learn to count better than conifers and pineapples and the trunks of palm trees and some camellia flowers, all of which give us the harmonious crisscrossing of 8 spirals with 13 others?

Perhaps you've already recognized that these numbers belong to the "Fibonacci series." In 1202, the mathematician Fibonacci counted a population of reproducing rabbits, which had started from a single couple. He found this series, of which each number is the sum of the two previous ones: starting with 1 then 1 again, we indeed get $1+1=2$, then $1+2=3$, $2+3=5$, $3+5=8$, $5+8=13$, then $21, 34, 55, 89, 144$, and so on.

But how can nature keep finding such specific numbers? It's pretty difficult to imagine that such mathematics, from the 13th century or otherwise, should be encoded in plant chromosomes. But then, aren't the forms of living beings a purely genetic affair? Is mathematics mixing with biology? By explaining how the Fibonacci series appears in the evolution of some dynamic systems, of which plants are only one example, Douady and Couder explained how simple physical processes in nature are the source of the selection of the forms of some living beings. That this discovery dates from 1992 shows that the problem, which has intrigued scientists for a century and a half, wasn't so easy to solve. Personally, such a challenge makes trying to understand it even more of a pleasure for me. Along the way, I'll treat you to a little history of science.

Leonardo Fibonacci's career was not ordinary. He was born in Pisa in 1170 and died there in 1250. He was the son of Guglielmo Bonacci, ambassador to North Africa of the merchants of Pisa, and Fibonacci was one of his nicknames. Since he found he didn't have the soul of a businessman, he also went by Leonardo Bigollo Pisano (literally, "Good-for-Nothing from Pisa"). Mr. Bonacci pushed his oddball son to learn mathematics from the Arabs, thinking it might be useful for accounting. Fibonacci stayed in the city that is now Béjaïa in Algeria until he was thirty years old, in contact with the best mathematicians of the time. There he came to understand the superiority of the decimal numbers and numeration that the Arabs had brought from India.

At the time, from Baghdad to North Africa, the great Arab scholars were inventing algebra, that is, formal computation, and many other things: think of all the words derived from Arabic, like algorithm, alembic, alchemy. When he returned to Pisa, Fibonacci, then, introduced Indo-Arabic figures and the system of decimal numeration into Europe, where they were definitively adopted; he thus put an end to the use of roman numerals. A revolution. Like his Arabic teachers, he would study the works of Euclid, and become famous across Europe for his own work, even getting a salary from

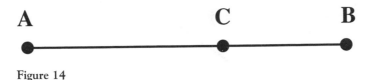

Figure 14

Frederick II, the German emperor who became king of both Sicily and Jerusalem, thanks to a crusade in those far-off lands (the sixth of its kind).

Interest in the Fibonacci series was revived four centuries later when Kepler related it to Euclid's Golden Ratio, which Fibonacci himself hadn't yet understood. The Golden Ratio is the proportion considered ideal by various painters and architects, ranging from Leonardo da Vinci to Le Corbusier, who claim to have found it in the pyramids of Egypt, the Parthenon in Athens, the form of the human body, of course, and so on.

Euclid, who may have lived at Alexandria from 325 to 265 BCE, or the name of whom may simply be the pseudonym of a group of mathematicians of the time,[39] introduced this ratio while looking for a harmonious way to divide a segment. We can choose a point C between the extremes A and B of the segment (see above), in such a way that AC/AB=BC/AC. The question of proportion seems to have previously preoccupied Pythagoras (born in 569 on Samos, a Greek island near Turkey, and deceased in 475, BCE of course), but Euclid is the one who found that this proportion equaled $(\sqrt{5} - 1)/2 = 0.618$. Then he noticed that it was also the relationship between the side of a regular convex pentagon and its diagonal (see below, EF/EG). It was by inscribing the "Vitruvius man" in a pentagon that Leonardo da Vinci showed that man had ideal proportions, with all the overtones that suggests. But Kepler discovered that the relation between two successive numbers of the Fibonacci series (3/5, then 5/8, then 8/13, et cetera.) tends toward the Golden Ratio as we advance along the sequence.

This number, then, has been a sort of symbol of rigor and beauty for twenty-five centuries.

I'm sure Douady and Couder knew all that, but they began with

137

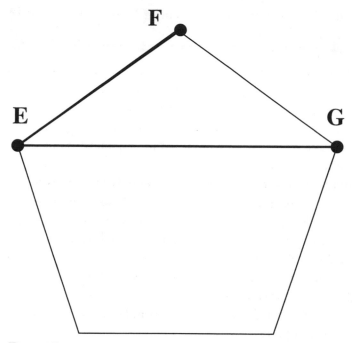

Figure 15

reading the work of the Bravais brothers. Auguste Bravais was a physicist, and ten years before he made his fame by classifying crystals, he became interested in plants. No doubt that was because his brother Louis was a botanist. It was together, then, in 1837 that they discovered that the Golden Ratio appeared in the shapes of plants (phyllotaxis). Douady and Couder's virtue is to have explained that the presence of the Fibonacci series and the Golden Ratio in phyllotaxis is a result of the physical conditions of the development of plant cells.

Like plants, flowers grow through the multiplication of cells around a point called an "apex." Next to the apex, at a certain distance, buds appear, which are the precursors of future leaves, petals, or scales (on a pinecone), or future sunflower seeds (called florets). As growth progresses, the apex moves forward, leaving the buds behind. But each new bud needs space in order to be created near the apex; it's as though its neighbors repel it at short distance. If the frequency with which the buds are being created is very slow, each new bud is repelled by just one other—the last one created—since growth has already taken the others far away. The new bud is therefore preoccupied with just the previous one, and appears opposite to it. The angle between two successive emerging buds around an apex is therefore 180 degrees: plants that grow slowly enough have alternating leaves on their stems. But if the frequency of emergence is a bit higher, each new bud feels the repulsion of several preceding buds, and the alternating configuration is not optimal: the buds shift themselves over in order to be as far as possible from several others. The angle between two successive buds is now less than 180 degrees. We can measure the angle as long as we note the order in which the buds appeared. The Bravais brothers discovered that the angle is close to 137.5 degrees, that is, 360 multiplied by 1 minus the Golden Ratio, which we call the Golden angle or section. Then, growth continues, and the buds get further away from the central apex, but their angles are preserved.

In order to look at how simple rules of growth could lead to these "Golden spirals," Douady and Couder first did a physics experiment. It consisted of raining down drops of a magnetic liquid in the middle of a container with some oil in it. Because of an exterior magnetic field, the drops repelled each other and were attracted to the outside of the container, which slowly brought them away from the center. The drops were therefore equivalent to the successive buds, and their slow movement towards the exterior of the oil bath corresponded to the displacement of the buds during the growth of the plant.

By letting the magnetic drops fall one by one, very slowly, Douady and Couder indeed found that each one went in the direction opposite to that of the preceding one. But by letting the drops rain down faster and faster, they saw that the drops arranged themselves in crisscrossing spirals: first 3 and 5 spirals, then 5 and 8, 8 and 13, and so on. As the frequency of the drops increased, the distance between each new drop and the preceding one diminished, the drops were packed together ever more closely, and were therefore

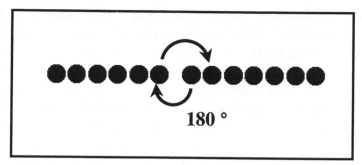

Figure 16

repelled by an increasing number of other drops; this forced the collection of drops deposited on the surface of the oil to make what Douady and Couder considered to be successive bifurcations between patterns including 3 and 5 spirals, 5 and 8, 8 and 13, and so on. Since this model of occupying space started from two simple physical rules—a repulsion between drops and a slow drift towards the exterior—it was easy to simulate on a computer. They therefore did a digital experiment in which they were able to extrapolate up to an infinitely fast rain. They demonstrated that what counted in the end for the selection of form was simply the interval between successive drops; they were able to continue the Fibonacci sequence out to infinity, and thus to find that the angle at which two successive drops appeared tended towards the Golden angle, that is 360 (1 -0.618) = 137.5 degrees.

In a commentary on these results that was published in 1995, Ian Stewart makes a remark[40] which seems interesting to me, since it makes it clearer why the Golden Ratio arises. If the buds appear by alternating from one side to the other, and the angle of their successive appearance is 180 degrees, we can see that the pattern they form will be two half-lines.

If the angle were a simple fraction of 360 degrees, like 1/3, that is, 120 degrees, the buds would be arranged as a star with three lines. With an angle equal to 360 * (3/8) = 135 degrees, we get a star with fifteen branches. In fact, for any angle equal to a fraction of two integers, we get a regular star, and at a far enough distance from the

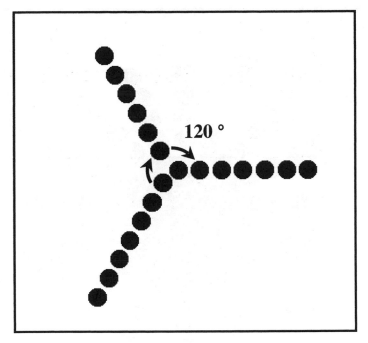

120°

Figure 17

center, this arrangement is not compact: empty spaces will appear between the buds away from the center.

To fill the available space better, it's preferable for the angle of appearance not to be what we call a rational number, that is, a ratio of two integers. And it so happens that the Golden number is the most irrational number—the most different from a rational number. All irrational numbers like $\sqrt{2}$, $\sqrt{5}$, π, or others can be obtained as the limit of a sequence of rational numbers, but we cannot find a sequence of integer fractions which tends quickly towards the Golden number. The successive ratios of the numbers in the Fibonacci series approach it, but slowly, and Ian Stewart explains thus that the patterns achieved by way of the Golden angle are especially different from the regular stars, and therefore the most compact. The figure below, constructed with the angle of the appearance of successive buds equal to the Golden angle clearly reveals a collection of

141

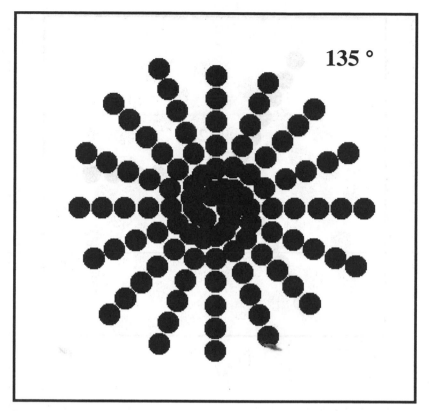

135 °

Figure 18

crisscrossing spirals that fills the space in a much more compact fashion than with an angle of 135 degrees, even though the angles are very close in terms of degrees.

So what might have appeared to be a magnificent, intriguing mystery of creation results in fact from the operation of two small, simple physical rules, and from the existence of a particular number, the Golden Ratio. Does biology therefore play no role at all? One piece of evidence that not every principle involved was biological, besides the fact that Douady and Couder's drops didn't have chromosomes, is that if we make an incision in the apex, we can force a plant which normally grows with alternating leaves to grow with leaves wrapping around in spirals. Conversely, biology must play

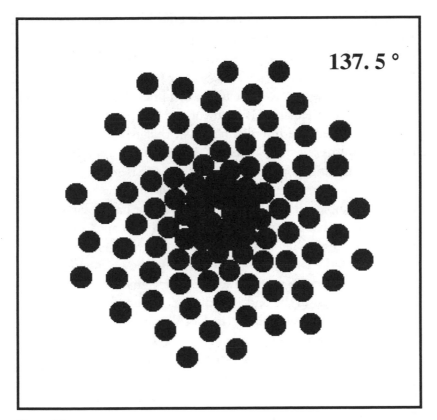

Figure 19

some role, since all cypresses make cones with 3 and 5 spirals, while no sunflower does that. According to Douady and Couder, biology only determines the rate of growth and the repulsion between buds, both of which are matters of chemical encoding in the chromosomes. Their great accomplishment, therefore, is to have relocated the problem of plant spirals within the general framework of dynamic systems, a very active domain in physics today.

In fact, the spontaneous organization of a dynamic system whose evolution in space and time obeys simple laws is a problem which has preoccupied scientists for a half-century. As early as 1952, for example, Alan Türing[41] thought that when you mix chemical products which both diffuse and react with one another, patterns can

143

spontaneously appear. Türing was thinking about plant spirals, but also zebra stripes or the spots on a leopard. The formation of spirals in chemical reactions would be discovered in 1990 by V. Castets and his team in Bordeaux. That's a research subject which is located at the intersection of at least four disciplines: biology, chemistry, physics, and mathematics. It is one of the best recent examples, to my mind, of genuine interdisciplinarity, that is, scientific progress which concerns several branches of science at the same time. Such research is getting more and more ambitious these days.

In like manner, some physicists are wondering how a protein or a long DNA molecule can fold itself to adopt a particular shape in three-dimensional space, rolled up like yarn into an X or Y shape, for instance, especially since that shape controls gene expression. The study of the causes of mad cow disease show that a biological molecule can become pathogenic if its shape changes. We are making rapid progress in the understanding of these forms: for example, we can recognize weak links in the DNA sequence which are easier to bend than others, and we hope to understand soon how the whole DNA molecule adopts its natural shape in our cells.

We are also wondering how an embryo develops. What is encoded, and what isn't? By pushing gently on a fly embryo at the very beginning of its development, we can cause two stomachs to develop instead of one; so, once again, not everything is encoded. During normal development, what leads to the appearance of an orientation to the body (there is a front, a head, and a back, a tail), then of segments, like a certain number of pairs of feet or rings among insects? Is it the same problem of the diffusion and reaction of chemical substances again? These are open questions, and current ones.

In my own way, by studying crystal faceting, I have also been working on the question of forms, ones which appear to be much simpler than some of the forms I've mentioned above. Investigating faceting has been difficult and interesting, for reasons related to the physics of surfaces, their fluctuations, and the role of thermal agitation. Nevertheless, when I think about it, by seeking to understand various forms, I think that I have been reasoning as a physicist, since I have sought to lay out a simple, general, and predictive model; but at bottom, I have also been trying to demystify the origin of the diversity of forms in nature—to replace the mysterious with simple physics.

I'm certainly not saying that the steps that allow an embryo to develop from a single cell to a human being with its head, its arms, its ability to do physics or to stay balanced on a bicycle will all be understood tomorrow—far from it. But science is rapidly progressing in this area, and no scientist today thinks of natural forms as impenetrable mysteries of creation.

From Pianos to the Sun

Enrico Fermi (1901–54) started his studies in Pisa. He was not only one of the great theorists of quantum statistical physics and artificial radioactivity, but also highly talented at performing experiments in nuclear physics. When he received the Nobel Prize in 1938 (he was 37!), he decided not to return to Italy, where his wife, of Jewish background, would be in danger. He emigrated to the United States and was named professor at Columbia University in New York in 1939. He pursued his research on radioactivity in Chicago and like many American physicists, participated in the development of the bombs dropped on Hiroshima and Nagasaki, through his work on the Manhattan Project.

Fermi liked to test his students' thinking abilities with questions like, "How many piano tuners are there in Chicago?" That's certainly a question which would baffle a few highschoolers in France today.

"That's not on the syllabus!"

"I don't know anything about that!"

Is it that hard? Let's think about it! How many people lived in Chicago in the late 1930s? Say, 3 million? And what proportion of them could have had a piano? Okay, 1%. So there would have been about 30,000 pianos in Chicago at that time; that seems reasonable.

And with those pianos being tuned once a year, that yields 30,000 tunings to be done per year. How many pianos must a tuner tune in a year to live decently? Let's transpose that to the present day: if a tuning brings in about 50 euros, he'll need about 40 per month in order to earn 2,000 euros; let's say, 500 a year? So in the end, the market for piano tunings in such a city should allow for 30,000 divided by 500, or about 60 piano tuners to live there.

That calculation is obviously crude. However, it shouldn't be completely off the mark. With the help of a few plausible hypotheses, we reached a number whose order of magnitude must be right. An order of magnitude is a range of which the upper limit is ten times as large as the lower limit. I would bet, therefore, that the true number of piano tuners in Chicago must be between 20 and 200. For cultural reasons, unfortunately, it has likely dropped since Fermi's time.

If we had supposed that a piano tuner takes 3 hours, including travel, to tune a piano, and works 36 hours a week, 50 weeks a year, we would have gotten 600 piano tunings per tuner per year, and so $30{,}000/600 = 50$ tuners in Chicago.

Objection: there are also concert pianos. How many concerts with pianos take place every evening in Chicago? Two? That would yield another 700 tunings per year, a minor correction compared to the 30,000 we started with. Even supposing there were 4 concert pianos to tune every evening, that would only make 1,400 a year, which would still be proportionately small—about 5%. Objection overruled.

That is how a physicist reasons. He (or she, though I'll stick with "he" in this paragraph, just for ease of discussion) starts by finding an order of magnitude for the phenomenon he's interested in. He then subjects his result to criticism. He checks that his approximations are justified. He confronts his reasoning with experimental results. He will want, for instance, to take a telephone book and check through a limited number of pages to see how many tuners he really finds, then do a simple proportional calculation after having checked the total number of pages of the phonebook.

But this problem about pianos is obviously simplistic. Here, we're only estimating the magnitude of a number, not testing a theory.

But that doesn't mean this approach is useless. On the contrary. For example: do windmills represent a viable source of energy for replacing oil, which heats up the planet, or nuclear energy, which produces waste we don't quite know what to do with? Let's think about it Fermi's way and take the example of France, which I know best.

A giant windmill spreads its blades across a diameter of about 100 meters. It can produce a maximum power of 2 million watts (2 MW). In order to do that, though, the wind can't be too weak, or the windmill won't turn. It can't be too strong either, or it will break. Moreover, the wind blows in gusts, not at a continuous strength. Let's suppose, then, that optimal conditions are attained a quarter of the time, thanks to a well-chosen location—perhaps the west coast of France, where the wind is abundant and relatively stable. Each windmill in Brittany would therefore provide about 0.5 MW. But there are about sixty nuclear power plants in France, which each produce about a GW (a gigawatt, that is, a billion watts). So French nuclear power plants produce a total of around 63 GW, 78% of the electricity in France. To replace all of these power plants, we would need at least 63 GW/0.5 MW = 126,000 windmills.

Whoa! 126,000 windmills at 100 meters in diameter would make a compact line 12,600 km long! Even building 1,260 km of giant propellers, which no coast-lover would be ready to accept, we wouldn't manage to replace a tenth of the nuclear power plants in France. In fact, we would be required to space the windmills out at least every 400 meters apart if we want to optimize their production, and so we would need 5,000 km of windmills to replace 6 nuclear power plants, producing only 8% of the electricity needed in France. Build a few to see the results, maybe, but I don't see them becoming an important source of energy. The wind itself really isn't powerful enough.

Calculating orders of magnitude is indispensable in many circumstances. So I wish every politician and environmental activist were in the habit of reasoning this way. And it's not simply a question of a state of mind. The future of energy on our planet is a complex problem, which requires good sense to begin with (hence my wind-mill example), but also the political and scientific courage to iden-tify the real problems and see how we might solve them, and without too many prejudices, even though everyone has an opinion on the matter. This is perhaps the time and place to speak about it a little.

Mankind finds itself between Scylla and Charybdis, like Ulysses. How should we choose between these two dangers? Should we keep burning fossil fuels, that is, oil, gas, and coal, upsetting our climate, or build more nuclear power plants, even though we don't precisely know yet what to do with the waste they produce? That seems to be the dilemma our society faces at the beginning of the 21st century. Even while some people have no idea that we need to solve this dif-ficult problem in the next few decades, others already have a firmly fixed opinion on the matter. Some speak of alternative energy; oth-ers are for nuclear energy, still others are resolutely against it. Faced with difficult scientific and technical problems in that complex de-bate, the physicists' duty is to inform the public and help look for solutions. With a few figures at hand, we can already understand a great deal.

World energy consumption is increasing constantly. In 2000, it reached 14,000 GW for 6 billion inhabitants, or about 2 kilowatts per inhabitant—though in fact it's much less in poor countries and much more in rich countries (5 kW per inhabitant in France, 11 kW in the United States, much less than 1 kW in underdevel-oped countries). Where do we get all that energy? About 32% comes from oil, 26% from coal, and 19% from gas, which adds up to 77% from non-renewable fossil fuels; the rest is split at about 5% from nuclear sources, 6% hydroelectric (dams), 10% from what's called biomass (mostly wood), and only 1 or 2% from other kinds of energy which are renewable, like solar and wind energy.[42] But there's no longer any doubt that burning fossil fuels has already led to a no-ticeable increase in the temperature of the Earth, because of impor-tant changes to the composition of the atmosphere. Even in France, where the figures are different from the world average since 39% of the total energy consumed (78% of the electricity) comes from nu-clear power, the problem of global warming is urgent. Thanks to

several satellites that are constantly observing what we are subject-
ing the planet to, the warming and its causes are now known with
certainty.

It's due to the "greenhouse effect," or more precisely to an increase in it. We put plants in greenhouses to warm them up. The
Sun's visible radiation comes in easily through the windowpanes (or
the plastic tarps), but has a hard time getting out. That's because inside the greenhouse, plants, people, the ground, and the other objects absorb sunlight, then re-emit the energy as infrared radiation.
But this invisible light can't pass through the windows, which are
opaque at that end of the spectrum. In the same way, the atmosphere lets in visible light from the Sun, which reaches the ground,
but doesn't let out infrared radiation, because of some of the gases
found in the atmosphere, like water vapor, carbon dioxide (CO_2),
methane (CH_4), et cetera. These are therefore called greenhouse
gases.

The greenhouse effect is very useful for life: without it, it would
be much colder on Earth. But the equilibrium is fragile: a bit more
CO_2, and the temperature rises. And we are burning more and more
coal, gas, and oil, which produces large quantities of CO_2. The CO_2
content of the atmosphere has doubled in 50 years, though it was
comparatively stable for 400,000 years. As a consequence, a notice-
able warming of the planet has already taken place: about 0.6 °C by
the year 2000. The warming might seem like little, but it seems inevitable that it will accelerate and reach at least 3 °C by 2100, under
the assumption that we considerably reduce our CO_2 emissions, and
maybe 6 °C if we don't manage to reduce those emissions in the very
near term.

And 6 degrees is a lot. Twenty thousand years ago, during the last
ice age, the temperature was 5 or 6 degrees lower, and the northern
portions of both Europe and America were entirely covered with an
icefield 2 to 3 km thick! (That had nothing to do with CO_2, but
rather with the Earth's movement around the Sun.) Sea level was
then about 120 meters lower than the current level! Of course, that
allowed us go from France to England on foot, but still, we can imagine what the reverse change—a warming of 6 °C—would do: melt-

ing ice, an increased exchange of water between the equator and the poles, a consequent increase in the number and intensity of extreme storms, changes to ocean currents, rising sea levels. A collection of major climatic phenomena would be unleashed by mankind over the course of a few decades, which is next to nothing on the scale of geologic time.

True, since there won't be much oil in forty years, CO_2 emissions will necessarily decrease. For a long time we were still discovering new oil fields, and the world reserves were constantly being revised upward. That's been over for a few years: we've looked in all the accessible places, and the reserves have started to diminish. Enough is left, in effect, to reach 2040 at present levels of consumption; most of it is in the Middle East, mainly in Iraq, Saudi Arabia, and Kuwait. One cannot avoid thinking that that is the underlying reason for the armed conflicts that have been started there. Oil will become more expensive as less of it is left, and clearly wars will only make the problem worse. A bit more gas and coal remain, but the problem of global warming is urgent; if we don't *immediately* reduce the consumption of fossil fuels, the amount of CO_2 in the atmosphere will continue to increase dramatically, and the consequences promise to be terrifying.

One might expect a physicist to favor nuclear power. Not a bit. I may as well state my position right away before explaining it: I'd like for us to be able to do without nuclear power, but I don't see how we can.

The security problems aren't really what worry me. The accident in 1979 at Three Mile Island, in the United States, didn't lead to the dispersal of any radioactive matter outside the damaged reactor. Only two people were irradiated, and the doses they received weren't fatal. To this day there has been only one serious accident: the one at Chernobyl, near Kiev. That was the disastrous consequence of an exceptional series of human errors regarding a reactor, the very design of which revealed serious security gaps. The reactor breach caused the deaths of a few dozen people who were at the site when the catastrophe occurred and around 2,000 cases of cancer which would cause the premature deaths of another few hundred people. It's difficult to estimate the number of those cases of cancer, since the threshold above which irradiation is really dangerous is poorly known. Those numbers are also based on the care the victims received. Some estimates go up to 20,000 potential cases of cancer in Ukraine, but these seem exaggerated. As tragic as it is, that accident should be compared with the rather high number of miners who die every year in coal mines, with the victims of dam breaches, and with explosions in chemical factories like the one in Bhopal in India (5,000 dead). Security problems in nuclear power plants are being taken into account to a greater and greater extent, both in how they are operated, as well as designed. Moreover, the justifiable demand for complete transparency of information regarding them has been satisfied. So I don't think that security is the most serious problem for nuclear energy nowadays.

The main problem is the waste. Among the waste, some is of low radioactivity, or has a short half-life. That includes, for example, the gloves, the filters, and various materials that have been contaminated. Their radioactivity fades rapidly, reaching natural levels of radioactivity after about 300 years. That is the A waste. All that's needed is to put it away and wait until it becomes harmless all by itself.

The B waste is of average radioactivity, and makes up about 3% of the total radioactivity produced in France up to now. At a rate of $20\,m^3$ per reactor per year, we would have accumulated about $40,000\,m^3$ of that B waste in 2004. Its cumulative radioactivity is high, but it releases very little heat, and scientists have been studying temporary burial methods in very deep layers of clay. That waste will also progressively return to the natural levels of radioactivity on Earth.

By contrast, the C waste is very active over very long periods of time. In particular, it includes plutonium 239, whose half-life is 24,000 years. It needs to be retreated. To do that, it is possible to reincorporate a part of that plutonium into the nuclear fuel, in the form of a combustible mix called MOX—but not all of it, and so the quantity of plutonium being stored is increasing. To store all that plutonium in the long term, scientists would need to master the technology of containers and storage sites so that they would last tens of thousands of years. It might be feasible, but that goal has not yet been reached.

Faced with these difficulties, would it be possible to recycle the C waste and operate on a constant amount of plutonium, which would have the added advantage of increasing the potential reserves of energy? That was one of the questions the fast-breeder nuclear reactor Superphénix could have answered, but it was shut down for political reasons. Reusing all the waste which, because it is recyclable, is no longer waste but new fuel, is nevertheless *the* solution. Reactors capable of burning their own heavy waste are among what have been called "fourth-generation reactors." They are still being studied, but might be put online by around 2050. They could allow us to provide electricity for tens of thousands of years.

I'm not claiming, however, that fourth-generation reactors don't pose any problems. I see at least two. The first is that recycling all of the plutonium 239 would require fast neutrons, which presupposes a special cooling system for the station. Thus, Superphénix wasn't cooled by water but by liquid sodium. Liquid sodium is a highly corrosive metal, and so handling it is difficult. Some have suggested using liquid lead for cooling, but that poses other problems. All of this means that more research needs to be undertaken and carried out. Another kind of problem is the quantity of plutonium in question. In effect, about 12 tons of plutonium are needed to get a 1 GW fast-breeder reactor going. That's a lot. For 60 reactors in France, that would mean 720 tons of this dangerous material, 6 times more than all of the French reactors have produced up to now. Other types of fourth-generation reactors are therefore being considered, which would use an element other than uranium 235: thorium 232, which is abundant on Earth. They wouldn't produce plutonium, would require less fuel to get going, and wouldn't need such high-energy neutrons. It may be possible to combine these different types of future reactors, then, and I think that if there is a future in nuclear energy, this is where it lies.

Of course, I only think it's important to accelerate research on fourth-generation nuclear fission because it would be a truly clean source of energy as regards both the climate and the waste produced, and because optimal use of the fuel would yield long-term energy, which is indispensable. Why indispensable? No matter how much we try to conserve energy, it will be extremely difficult to stop consumption from increasing, especially since many countries are justifiably asserting their right to develop. Under optimistic scenarios, world energy consumption will have doubled by 2050. If we want to cut CO_2 emissions in half, the proportion of fossil fuels must therefore by reduced to a quarter of the total, that is, it must go from 77% to less than 20%. That strikes me as impossible if we have to do it without nuclear energy. But let's consider other possible solutions.

People often speak of thermonuclear fusion, as though it were a clean and inexhaustible source of energy. In his plea for the experimental ITER reactor, which Europe, with support from Russia, China, and South Korea, succeeded in having built in Cadarache, even though the United States and Japan were pushing for the Japanese site of Rokkasho-Mura, Jean-Pierre Raffarin (the French Prime Minister at that time) declared on November 17, 2003, "This is the project [which will bring us] the energy of the future, practically inexhaustible and without significant harm, thanks to the abundant resource of hydrogen contained in water." Thus, by "mastering the energy of the stars," future thermonuclear fusion plants would solve our problem. His words suggested that the solution was as good as our friend the Sun.

Sadly, that's not at all the case. In a hypothetical "fusion" reactor, nuclear reactions produce helium from hydrogen, whereas in a classic "fission" reactor heavy elements (uranium, plutonium, thorium) are broken apart. Helium is indeed a perfectly harmless gas, but the fuel is not ordinary hydrogen, abundant everywhere in the form of water. It's a mixture of its two heavy isotopes, deuterium and especially tritium, which are anything but easy to make in large quantities. The problems for any future use of thermonuclear fusion are very difficult, and for the moment not at all solved.

As we can easily imagine, sticking a little Sun or a fraction of an H-bomb in a sealed box is not a simple plumbing problem, even though the directors of the ITER project didn't mention it much. The fusion reaction produces neutrons of colossal energy—14 million electron-volts (14 MeV)—whereas even in fast-breeder reactors the average energy of the neutrons is about 1 MeV. Subjected to a flow of 14 MeV neutrons, the steel of the surrounding walls would get a serious working-over. In effect, in the internal walls of a fusion reactor, not only would the atoms suffer so many collisions that they would each have been moved more than a hundred times within five years, not only would this lead to a swelling of the steel, which would stand little chance of remaining sealed after that kind of internal rowdiness, but the radioactivity created in the steel would produce "alpha" particles, that is, helium nuclei. That would lead to actual boiling, unless the helium is released. But how do you invent a material that is both porous, letting the helium escape, and absolutely impermeable, in order to keep the deuterium and tritium mixture, which must be heated to 200 million degrees, in an ultrahigh vacuum? To my knowledge, no one has any idea.

As for tritium, it's a gas which up to now has been produced only in small quantities for H-bombs, and serious production of it from lithium would require a thorough preliminary study. The plan is to circulate lithium inside a double wall, the point of which would be to produce tritium under irradiation and to protect the exterior wall at the same time. But the experimental ITER reactor won't be testing lithium production; it's only built to test the flow of a gaseous mix of ionized deuterium and tritium, a sort of big circular lightning bolt inside a giant ring-shaped magnet—what's called "the magnetic confinement of a hot plasma." So it won't answer the prior questions about fusion that will determine its future as a commercial source of energy.

So fusion isn't just around the corner; at most, we'll know by 2050 if it's a model for the future. Some people think now that building the experimental ITER reactor was a huge waste (5 billion euros for the construction and at least as much again for running it), as it only interests a few physicists in the field of plasmas, and it now appears completely premature as regards commercial energy production.

Many of our Spanish, American, and Japanese colleagues are congratulating themselves that their countries weren't chosen for the site! The giant machine goes on the long list of dubious decisions made according to the maxim shared by some technologists: "Big is beautiful." Personally, I compare the project to the International Space Station (100 billion dollars!), which is absolutely useless, despite the fact that our ex-Minister of Research, Mme Claudie Haigneré, stopped by for a brief stay. Robots are better for observing the Earth or various other objects in our local Universe, since they're stable and immobile, whereas astronauts upset the stability of the machines, need food, need to move around, need to get up there and back down again and be protected from all the risks they run by going up. In fact, practically all the experiments the astronauts up there are actually involved in are concerned with their own health. But what's the point of studying astronaut health in the dangers of space (radiation, weightlessness) if sending them into such conditions serves no other purpose? Similarly, why study the stability of a deuterium-tritium plasma at 200 million degrees if we'll never know how to place it in the right kind of sealed box to produce electricity in any case? Grandiose projects aren't necessarily part of good research policy.

Given the problems with both oil and nuclear power, why not seek out other sources of energy? The main available source is negative and very important: it consists simply in all of the savings we could make. The potential progress in this domain is considerable, whether in lighting, thermal insulation, automobile transportation, or industrial production. Let's stop using our cars to go 500 yards or using an airplane to jump over puddles! Let's stop wasting all sorts of disposable products whose manufacture takes enormous energy! Let's insulate our houses and save on heating in the winter and air conditioning in the summer! Let's stop eating artificially ripened strawberries out of season! This is a moral issue on a global scale. Nonetheless, while rich countries should and can stop wasting energy and resources, poor countries have a legitimate right to use more. By 2007, the total CO_2 emission of China became comparable to that of the United States. But the emission per capita at the time of writing was at least 4 times less in China than in the United States. The problem is clearly not merely technical: it's political, human, and global. So it's not very likely that we'll manage to noticeably reduce the total consumption of the planet. But we need to cut emissions of greenhouse gases at least in half, and quickly, if we want to avoid upsetting the climate.

Are there other sources of energy available? There isn't much free space for hydroelectric dams, and that's a pity. China is nonetheless getting ready to build enormous hydroelectric installations. They will, I hope, be careful about the health risks represented by hundreds of square kilometers of stationary water.

We would be wrong to ignore research on windmills and the ways they can complement our other sources, but that isn't where the solution lies. It's not so easy!

I do believe that solar energy has some potential. Let's look at the numbers in France again. We receive an average of 100 to 200 watts per square meter from the Sun, depending on the latitude. That's an average which takes into account absorption by the atmosphere, alternation between day and night, good and bad weather, and so on. Still, that's a lot: 100 to 200 watts per square meter. If everyone living in France could collect the energy available over roughly 30 square meters, they would provide their total consumption right there. Unfortunately there is a lot of loss, especially if you want to transform solar energy into electrical energy, not to mention the fact that it's often necessary to store that energy before using it and that current batteries are not good enough for that. It's easy to use solar energy to heat water by running it through black pipes, and this thermal solar energy source should be immediately developed everywhere. The use of solar energy to produce electricity is more difficult.

The maximum theoretical output of a silicon solar cell is 25%. The best collectors created up to now, with monocrystalline silicon, are expensive, difficult to manufacture, and have an output efficiency of just 12%. Polycrystalline silicon collectors are cheaper and have an output of 10%. Does the future lie in semiconductor films that can be incorporated into construction materials? They are

157

already available as of now, and their cost would obviously drop further if they were produced on a large scale. But their actual output is only 5%. And watch out for the energy cost of manufacturing them! To produce 30 GW of electrical power, about 10% of the total power consumed in France, you would need to cover at least 4,000 km^2 with collectors of this type—almost 1% of the territory, or 60 m^2 per inhabitant. That's already more than the total surface area of all the roofs in France, to supply only 10% of our needs. But regular and judicious use of solar energy could yield a useful complement to other sources of energy, especially if we make further progress in developing solar cells and storage devices.

A few words about biomass, that is, wood and various agricultural alcohols. They are only a renewable source of energy if we replant as much as we consume. As the new plants grow, they capture the carbon from the CO_2 released when the old ones are burned. Moreover, wood is clearly easy to store. Unfortunately, it's necessary to point out that the energy available from the plants we burn, wood or otherwise, comes originally from the Sun: it's solar radiation that permits photosynthesis, that is, the storage of energy in the form of wood, which is then burned. But the output of photosynthesis is very low: less than 0.5%. That means that for one watt of power received from sunlight, a field only captures 5 milliwatts. Nature, for once, is less than impressive. Even the cultivation of sugar yields only 0.6 W/m^2. So, in order to serve the energy needs of the planet, we would need to dedicate to biofuels a surface area about 20 to 100 times larger than that currently being used for agricultural purposes. What's more, we would need to take into account the fuel consumption of agricultural machines! Biomass will never be more than a complementary source of energy. What's more, developing such an avenue would run the risk of large-scale emissions of methane, a much more impressive greenhouse gas than CO_2, or the massive use of manure, which would put the quality of water at risk—another current planetary problem.

The energy problem, then, is serious, difficult, and urgent. Producing energy is crucial, unless we're willing to go back to the time when we only utilized slave labor—but a slave yielded barely more than 100 W (enough for one light bulb)! And the problem doesn't end at production: we need to store and transport the energy. The success of oil surely comes from the fact that, even though it's highly flammable, it's an easily transported source of energy. Post-oil (and that starts now), we are looking at the widespread use of what are called "fuel cells." If you submerge two electrical wires into water and you run a current, hydrogen will be released on one side and oxygen on the other: you have electrolyzed the water. A fuel cell does the reverse: it consumes hydrogen, producing water and electricity. Such cells were invented to produce electricity in spaceships, and today there are several kinds. Small ones are used to power mobile phones, larger ones for cars, and there are even larger ones than that (up to 10 MW). They don't produce any CO_2 as long as they are truly running on hydrogen rather than alcohol, as some do. We could thus make hydrogen from nuclear or solar energy, and then perhaps mount that hydrogen on cars, buses, and airplanes of the future, but hydrogen is not easy to store. We should also be able to hydrogenate biofuels in order to greatly increase their energy potential. Fuel cells might transform our means of transportation over the next few decades. It will depend on the inevitable increase in the cost of oil production and the international agreements which should be signed with the urgent aim of reducing CO_2 emissions. It's becoming crucial that the United States agree to cooperate with the Europeans, Russians, and Japanese in saving the climate, by signing the Kyoto treaty, which is only a preliminary step on the hard road which we are facing.

The consumption and production, the storage and transportation, of energy is a major problem. Its solution is only partly known and requires a considerable amount of scientific and technical research. It also demands public debate in as cool-headed a manner as possible. Fermi-style reasoning and calculations in terms of orders of magnitude aren't enough to solve the technical problems. Still, they strike me as indispensable for distinguishing the avenues which are genuinely interesting from those that are based on acts of faith, intimidation by economic or political pressure groups, and sheer demagoguery.

I Speak English

"It is regrettable that pseudo-researchers think themselves more interesting when expressing themselves in English!"

Thus spoke Mr. Jean-Louis Masson, a conservative member of parliament from Moselle, during the session of Wednesday, May 4, 1994, at the French National Assembly. I don't think he was targeting me personally, even though he might remember—unless I'm mistaken—that we were in some classes together during our studies. Still, though, what a tragic lack of understanding!

When I entered the sixth grade, in October 1957, my parents made me choose German as my first foreign language. Was the idea to cement a reconciliation with that country, after a fanatical minority, allied with the Vichy police, had dragged my grandfather in 1942 from the Vélodrome d'Hiver first to Drancy and then to the gas chambers at Auschwitz? At the end of the school year, a family in North Rhine-Westphalia was found to host me for a language sojourn. I was ahead in my studies and was thus not yet eleven years old when I found myself at the home of Herr B., a schoolteacher who practiced corporal punishment in the classroom. I can still remember his cane. With pride and feeling, he showed me the beautiful photo album which glorified the occupation of Brittany by his division of the Wehrmacht.

160

Herr B. also asked me about my family; whether, for instance, my grandparents were still living.

My parents had raised me in near-total ignorance of the extermination of the Jews and how that related to our family. No doubt they wanted to protect us. Or perhaps their memories were unbearable. I did know that my paternal grandparents had died, but I didn't know how. I also knew that my grandfather on the Balibar side had been born near Kiev, and that, disappointed in not getting French citizenship, he would have liked to return there. Hesitating slightly, I responded therefore that my grandfather died somewhere "between France and his native Russia." It was true: Auschwitz was indeed somewhere in there!

I still wonder how long it took Herr B. to realize what the background of my father's side of the family was. Despite the circumstances, of which I remained blissfully unaware at the time, I learned German every day. Six weeks later, I spoke it fluently.

When I graduated into the next year, I was horribly bored in German class. None of the other students had gone to Germany, and very few were able to string even a few words together. As for the teacher: believe it or not, but it's true, he made us listen to the Nazi anthem. The Horst Wessel Lied in the middle of class! I prefer to leave out his name. Subjected to such boredom, over the course of a year I forgot all the German I had learned in those six weeks. During high school, I had to learn it again during another stay in Germany, and for lack of practice, forgot it all over again later. I never learned to speak it fluently again. I still have the desire to support Franco-German friendship. I may also say that when I accompanied, on the piano, a friend of mine who sang Schumann's *Dichterliebe*, my knowledge of German was still sufficient for me to understand Heine's poetry and to try to adapt my playing to the memories of past loves the poem evokes.

Regardless, my German education in high school was largely a failure. The same causes led to the same effects when I attempted to learn Russian in junior high school. I took great pleasure in reading Pushkin and Gogol in the original, but I never managed to express myself with ease, and it all evaporated even more quickly afterwards.

I still know the alphabet, which helped me find my way around Moscow in later years; that's better than nothing, I suppose.

Today, I'm often crisscrossing the globe to explain my scientific research, and the more I hear scientists speaking in English, the more I've come to the conclusion that on average the French are pretty bad at foreign languages. I think only the Russians and the Japanese are worse. And, let's be realistic, unless it's presented in a way that makes it easily accessible on a global scale, a piece of scientific work doesn't have a chance of achieving recognition. What's the point of all that sweat and hard work to move knowledge forward if no one will find out? Moreover, without easy access to the work of his competitors, a researcher is sure to lose the hard competition in which he's engaged. Why, then, are the French so much worse at English than the Germans or the Dutch? During my brief time at the *Conseil National des Programmes*, we asked the same question. I did the following calculation.

How many hours a year does a student spend speaking the language she's chosen to learn in junior high school? Let's suppose that the language is English, and that at a rate of four hours a week and 35 weeks of class per year, she has 140 hours of English classes per year. In that class, around 35 students will share the floor equally. In addition, there will be written exercises, and the teacher speaks as well, so the students speak for only a quarter of the time. In the end, each student speaks only $140/(4 \times 35) = 1$ hour per year! Over the course of seven years of schooling, that will still add up to less than a day.

That's astonishing, isn't it?

But isn't a young Dutch student in the same situation? Of course. Some people think that if we did English in primary school, young French students would do notably better. But even if we all agree that the brain is more malleable at age eight than at age twelve, the same calculation shows that such an effort would be in vain: a child won't make real progress speaking an hour of English per year. I asked my Dutch colleagues how they learned English, whether they started younger (they don't) and how many hours of it they did at school (about the same as in France).

"But, in Holland, the children know English *before* they study it at school! All they need to do is watch television!"

That's how I learned the obvious: if we subtitle foreign shows instead of dubbing them, we not only respect the work of the director, we also teach our children languages. Having mastered the basics of the English language, Dutch children perfect it at school and study literature. They don't need to start taking classes at eight years old to do that. Simple, isn't it? I have a Tunisian nephew who learned Italian that way, and now he speaks four languages fluently.

With just a little German and a bit of Russian, then, I found myself at age twenty-two confronting an article in *Physical Review*, the pre-eminent American physics journal. Fifteen compact pages of scientific English. I still had some illusions about my intelligence then, and that task brutally gave me the modesty I was missing. Understanding a single article was going to take me a considerable amount of time. But I could not reinvent all of contemporary physics by myself. So I picked up a dictionary and suffered in silence.

I bore my handicap as best I could for five years, the time needed to obtain my doctorate, and then hurried off to England. I wanted to further my research on quantum evaporation, and take advantage of the location to learn English. There in the country of Robin Hood and Lord Byron, my immersion was complete and effective. By listening to and repeating what people said to me, within a few weeks I was able to get by in an ordinary conversation and basic scientific English no longer posed a real problem for me. That visit saved me, and since then I haven't stopped using my English, no doubt with some grammatical gaps, but also with the great satisfaction of being able to express myself painlessly in scientific discourse, and of understanding whatever my colleagues say to me, whether about science or everyday life.

I can no longer live without my international communication tool, English. I wouldn't give it up for anything: it's a matter of freedom, on which I'm unwilling to compromise. How, though, did Mr. Masson think that scientists actually work? Has he ever found himself at a conference bringing together participants from at least ten different countries? If so, what did he do? Scientific research has been globalized for a long time. The boom in means of communication, airborne and electronic, has shrunk the planet. The competition is therefore livelier than ever, and not understanding what others have discovered, or failing to make others understand what you yourself have found out, is equivalent to scientific suicide. Even in the past, when the world seemed bigger, there was always a need for a language of international communication. Depending on the time and the region, it may have been Chinese, Greek, Hebrew, Arabic, Latin, or French, but today, it's English. Being dismayed at American hegemony or screaming in defense of the French language won't change that in the least.

I was therefore flabbergasted when, in early 1994, Jacques Toubon, the Minister of Culture and of the French Language, wanted to require researchers to speak and write in French under various circumstances in their professional activities. It was during a discussion of Toubon's law[43] that Mr. Masson spoke up. It was a stormy session. In the version which passed, the law specified:

Article 6.

Any participant in a presentation or conference organized in France by natural or corporate persons of French nationality has the right to express himself in French. Documents distributed to the participants before and during the meeting to present its program must be written in French and may contain translations in one or several foreign languages.

When a presentation, a colloquium, or a conference involves the distribution to the participants of preparatory or work documents, or the publication of acts or proceedings, the texts or presentations given in foreign languages must be accompanied by at least a summary in French. These arrangements are not applicable to presentations, colloquia, or conferences which concern only foreigners, nor to presentations for the promotion of commerce outside of France. When a corporate person in public law or a corporate person in private law charged with a mission of public service initiates presentations directed at the present article, a mechanism for translation must be put in place.

Article 7

Publications, journals, and communications distributed in France and which emanate from a corporate person in public law, a private person executing a mission of public service, or a private person benefiting from public subsidy must, when written in a foreign language, include at least a summary in French.

The granting by a public person of any aid to works of teaching or research is contingent upon the commitment made by the beneficiaries to guarantee publication or distribution of their work in French, or to execute a translation into French of the publications in foreign languages which they involve, unless dispensation is granted by the minister of Research.

The ending of article 7, in italics, was thankfully eliminated by the Constitutional Council. Without that exclusion, it would have become impossible for us to publish the results of our research in English unless we also published them in French. Unless we agreed to be read by none of our foreign colleagues or competitors, we would have had to write up our results in duplicate: a French version and an international version. Taking into account the brutal nature of competition in research, that handicap would have killed research in France. The Constitutional Council, however, isn't supposed to make judgments with the protection of research in mind but merely as to the coherence of the laws. It was therefore because this ban represented a hindrance to the freedom of expression, which is firmly asserted in the Declaration of the Rights of Man and of the Citizen, the preamble to our constitution, that the Council eliminated the last paragraph of article 7 of Toubon's law. The rest all passed, and in theory is in effect today.

What does all of this mean in practical terms? There are several levels of publication, depending on whether we're dealing with original results from work in progress, which we call our "primary publications," or review articles, books for teaching, or articles for a wider public than our specialist colleagues. Regarding our primary publications, a French journal in principle does not have the right to a public subsidy unless it attaches a summary in French to each article published in English.[44] If it had been applied, this constraint would have quickly dissuaded any foreign researcher from publishing in France. In effect, how would our colleagues have been able to verify the exactness of the translation (which would have been done by whom?), unless they had learned French? Wouldn't they have considered it much simpler in these circumstances to publish elsewhere? Specialized French journals would have quickly disappeared from the market. I know of no international journal of physics which requires a bilingual summary from its authors.

In fact, history sometimes marches faster than the work of the National Assembly, and the problem never came up. With the European Union moving forward, the need and the advantage of collecting different national journals into larger European journals were quickly felt. In physics at least,[45] the French, Germans, and Italians rapidly decided to combine their national journals, which became European and thus have no truck with Toubon's law.

However, the article dealing with conferences was not censured. The legal obstacles to holding international conferences in France are therefore legend. In effect, unless we finance these conferences without the least bit of help from the State, all or most documents involved must be bilingual. The conference announcement must be bilingual, the program, the collection of summaries we distribute at the beginning of the conference in order to know who is speaking about what in an assemblage which often brings together more than a thousand participants speaking in parallel sessions in several different rooms. Forget about the cost of paper such a constraint would impose. Let's also overlook the considerable amount of work the translation of thousands of abstracts would be for the French organizers. Let's be content to realize that you can't publish a French text with an author's signature of approval if he or she doesn't understand it. Finally, the debates themselves must, in theory, be furnished with a "mechanism for translation." To understand what that means, it's best to take a look at the minutes of the debates in the National Assembly.

It was, as I mentioned previously, a stormy session, despite a few lucid moments like this one:

Mr. Bardet: If you require that French be spoken at any conference taking place in France, no foreign presenter will come! And if they can use whatever language they want, as Mr. Brunhes suggests, conferences will become towers of Babel!

The primary nationalistic point of view won the day, after a debate in which one could hear remarks such as the following:

Mr. Sarre: Scientists should be the first people to defend our language. If they don't do it, it's for purely careerist reasons. The more articles they publish in English, the better their careers will be.

Mr. Masson: It's unthinkable that a French researcher, paid by public funds, should express himself in France in a language other than French.

Mr. Jean-Louis Beaumont: You are correct!

Mr. Weber: How can we prevent that without imposing punishments?

Mr. Fréville and Mr. Leroy: Good point!

Mr. Toubon: The punishment appears in article 13: the loss of subsidies.

When the application decrees for Toubon's law were published on September 6, 1995, the threat of sanctions became explicit. In their article 17, fines of up to 5,000 francs per infraction were prescribed,[46] 25,000 francs for an organizer. And if anyone opposed an agent coming to record an infraction, it could go up to 50,000 francs, along with six months in prison! Moreover, five associations for the defense of the French language were authorized to prosecute us before the courts (article 19) in case of failure to observe the law.

The discussion about arranging for a "mechanism of translation" will illustrate the contradictions in which the minister was entangling himself. Mr. Jacques Brunhes demanded simultaneous translation. The cost of such a translation, which he estimated, per day and per language, at 8,000 francs for the booth and a 3,000 franc honorarium for the interpreter, seemed to him to be "not dissuasive." Did he have the slightest notion of the budgets we work with?

Mr. Mathus insisted, "What would be the point of consecutive translation? Are we going to, once again, legislate for nothing? Why should we include inapplicable mechanisms in the law?"

Mr. Toubon tried to defend the idea of a "consecutive" translation which would be "instantaneous," following the model in which, according to him, interpreters intervened in discussions between the Russian and French presidents of the time, Mitterand and Yeltsin. In the end, the Assembly adopted the amendment proposed by Mr. Fanton: "A mechanism for translation must be put in place."

Despite our joking in the lab about the possibility that "redox potential" could be translated as "red ox potential" by some interpreter who would doubtless be ignorant of oxidation/reduction reactions in chemistry, the situation was really serious. Even though the law was formulated with a certain degree of vagueness, we were feeling anxious. How could we force foreign speakers to write abstracts in French? Was some pro-"linguistic purification"[47] association going to prosecute us if we didn't translate? Would a pocket dictionary left at the main desk of the conference constitute a "mechanism for translation?" Were we going to have to back out of doing our share in organizing international conferences with the excuse that some legislators in our country had no idea of how our profession is practiced?

Our hackles were raised, and Mr. Mathus was ironically proven right by the way things unfolded: in the world of research, the law was "totally inapplicable" and "grotesque," and the Assembly had indeed "legislated for nothing." I obviously don't mean that we should have accepted his demand for simultaneous translation in all conferences in France. I'm just happy that Toubon's law was never enforced, and everyone proceeded to forget about it. I also hope this book won't serve to bring it back from the dead!

Does that mean I'm not interested in the defense of the French language? Certainly not. As I wrote at the time,[48] by learning English, French scientists would acquire the means to keep their place in the world. That's the price we must pay for attracting foreigners to France, where, once they've arrived, they will no doubt take great pleasure in speaking French, absorbing French culture, and who knows, maybe even doing their scientific thinking in French after a few weeks of total immersion. What I propose, therefore, is the exact opposite of drawing Maginot Lines (the French defense against Germany before the First World War); it's rather to take France, its territory and its scientific culture, and open it up to the rest of the world. The Soviet empire built its iron curtain, forcing scientists to publish in Russian and exercising strict control over their comings and goings. The result was obviously catastrophic. Soviet science collapsed along with everything else, and the best scientists departed en masse to the West the moment the revolution of 1989 began. Instead of paying interpreters who don't know the relevant specialized terms anyway, and whose translations are inevitably filled with unacceptable errors, we would do better to offer language instruction to all the foreigners we draw to France because of the high quality of our work. That would defend the French language in a different way—a much better one. Nevertheless, I don't think that will be enough.

Just as we need to preserve the diversity of species in order to benefit from their adaptive possibilities, so I think that preserving multilingualism in science is likely to enrich its possibilities. Haven't we all noticed that we change our personalities slightly when we switch languages? English seems more concise and more supple, which may have helped it acquire its status as the language of international communication, but the political and economic dominance of the United States is clearly a much more salient cause of the present state of linguistic affairs.

French has, nonetheless, certain cards to play. Are some languages better suited to rhetoric than others? In a talk in which he expressed a point of view similar to my own,[49] Roger Balian pointed out that the French conjunction *or* doesn't exist in English. Regardless, it's clear that truly bilingual scientists are very rare, and that it would be difficult for French scientists to express themselves as precisely in English as they can in French. So I'm keen to keep doing my thinking about physics in French, but it's not so easy.

It's certainly impossible without the vocabulary. But as new concepts appear and theories are refined, and even as some metaphors are adopted as illustrations of that progress, science invents words and dedicates certain phrases for certain uses. If we were happy just to participate in this linguistic evolution of English, scientific French would become a dead language. It is therefore crucial to translate those words and adapt those phrases, and so to keep scientific French alive. How should we do that?

Helping us translate the basic textbooks into French would be good, so that we could keep teaching in French. In the countries with which I am most familiar, like Italy, Finland, and Holland, it seems that a large part of higher education has switched to English; but I'm not ready to accept that we've arrived at that point in France. By translating, we will be forced constantly to make the meanings of our words precise. The French language will improve, and we will turn the handicap of not having the language of international communication as our native language into an advantage.

We should also make it a habit to disseminate scientific information in French in newspapers and books for non-specialists. Regaining a bit of subtlety of expression by talking about physics in French is a pleasure to which I am just as attached as I am to that of making myself understood in English before a multilingual audience. And there again I find it healthy to pay attention to the precise meanings of words, to their primary senses as well as their connotative powers.

Finally, there are still circumstances, though they are becoming rare, under which our debates are specialized but do not require English: internal seminars, French or bilateral work meetings, exam sessions, or evaluations of our activities. Why rare? Because evaluations of our laboratory are being performed more and more by international juries. And because even the smallest work meeting at the national level always attracts foreigners, among whom we should, out of politeness, express ourselves so that we can be understood.

It isn't really that ambitious or unrealistic to suggest that every French student should learn not only English, the international language, but also an additional foreign language. To that end, understandably, I suggest widening the practice of language sojourns rather than expecting a student to learn English in just one hour of class time per week in elementary school. I also suggest subtitling foreign films and television shows rather than dubbing them. Daily training in switching between languages is an especially stimulating intellectual exercise. Don't the young bilingual students we happen to know tend, often, to be the brightest? Don't our brains need as much daily exercise as our muscles? And isn't the practice of multi-lingualism one of the best possible vaccines against xenophobia? Provided that the intellectual heirs of Messrs. Toubon, Masson, Sarre, Fréville, and Leroy don't make our civilization regress!

What Don't I Know?

My mother was interested in the birth of the French language in the ninth century, when Charlemagne's grandchildren were splitting up his empire, as well as the birth of French literature not long after. Two of the books she wrote were for the "What Do I Know?" series of books, published by the Presses Universitaires de France.[50] Now when I change the question a bit, to "What *don't* I know?" I want to show that I share with her a certain taste for knowledge, but with a difference: I prefer, in effect, the uncertainties of research to the certainties of taught science.

Don't misunderstand me: I neglect the pleasures of acquired knowledge only because I want more of it; I'm more interested in advancing its frontiers. In passing, let me make clear that I wouldn't think of reducing all knowledge to a temporary state of ideas, as though science fluctuated without making progress. No, science moves forward, inexorably, and I always find the way it advances fascinating, with its ruptures followed by periods of deepening understanding. Besides, by displaying my taste for the evolution of science and the frontiers of knowledge, I remain my mother's heir: her intellectual activity was rich, and she didn't shy away, in her other books, from putting forward ideas which were powerful enough to shake up the subject she was studying.

172

By admitting to a certain amount of ignorance, I'm also taking the risk that someone might doubt my sincerity. True, I wouldn't dare to admit ignorance about questions I really know nothing about. I reveal my ignorance only in those areas where, on the contrary, I know enough to know that I'm not the only one who is ignorant—even among those who could claim to be experts on the matter. Speaking from my perspective as a physicist and disclosing what I don't understand about physics therefore presupposes a certain level of knowledge, without which ridicule would be guaranteed. Have I avoided that in this book? Only time will tell, but in any case, that isn't what is most important to me.

What I want most is not to pretend to know, not to overcome any power or position by force, not to claim that science is an unassailable edifice from whose heights we scientists can reveal the truth to the ignorant, who need only accept it. I'm ignorant too, just like everyone else is! That won't be changed by the fact that I want to share with my readers the excitement of learning for myself that the world is not the way we thought it was, even a few years ago.

The world is, indeed, not like we believed it to be in 1950. Even Einstein hadn't understood at that time that the Universe is unstable. Does his ignorance at that time undermine his reputation, or his status as a symbol of absolute genius?

Scientific truth makes progress. The fact that science questions itself doesn't mean that scientific truth has no value, rather, it's the opposite. By definition, a researcher wonders every day whether he is wrong, and it is that which, in my view, makes for both the immense difficulty and the greatness of the profession. By relying on the rigor of a method that constantly calls for modesty, the scientific researcher has to be able to convince others with arguments; he should never have to impose any statement by force. No intimidation; only the power of reason!

So basically, I'm a scientist fighting the "worship" of science. Unfortunately, in a society that doesn't always take time to think, scientists are often tempted to assert a kind of authority over the general public on the basis of having mastered a language whose terms are inaccessible to the majority. Is it really fair to extend the authority acquired in one domain (the scientific domain in this case) into another? Does mastering all the details of a difficult body of theoretical knowledge such as quantum physics or statistical physics qualify us to claim that we hold the truth about genetically modified plants, the ethics of science, or the morality of industrial development? If it were only a question of modestly transferring a rigorous method of thinking into an area we know we poorly understand, such claims might be justified. But, even then, we should be cautious.

At the same time, I must admit that I'm disturbed when nonscientists pretend to master esoteric vocabulary and dress up their talks with scientific adornments in order to seem wise; they contribute to the image of science I'm trying to resist. Instead of stimulating people to engage in independent thought, instead of developing their audience's sense of critical awareness, they foist their claims upon them, they indoctrinate, they abuse their power through the use of jargon. That's exactly the opposite of what I try to do.

Some French philosophers or psychoanalysts have been prolific in this kind of perversion. For an especially enlightening analysis of this kind of writing, I suggest reading Alan Sokal and Jean Bricmont's book *Fashionable Nonsense*. In that excellent book, the quotes are abundant, never truncated, and pseudoscience is unmasked. I should add that Sokal and Bricmont attack nonsense itself, not its purveyors.

In recent French literature, one of the most striking examples is Michel Houellebecq, a highly successful writer who deliberately presents himself as a prodigy, and as having mastered all of physics and mathematics, before moving on to literature. I have a hard time believing he's sincere about his intellectual past. I think, rather, that it's a game he plays, following a recipe for commercial success. His writings are worth a glance, at least to use as a foil against the works of true scientists, as well as earnest popularizers of science.

In his novel, *Atomised*,[51] for example, he writes about an article published by his protagonist Djerzinski: *"Three conjectures on Topology in Hilbert Spaces*, published in 2004, was a shock. Some interpreted it as a reaction against a continuous dynamic, as an attempt—with strangely Platonic resonances—to reestablish an algebra of forms."

Imagine my surprise when a friend's daughter, studying comparative literature in New York, asked me what this meant, since I was someone who must understand modern physics. The illusion of profundity created by this gibberish works so well that they are examining its possible meanings in American universities!

In his earlier novel, *Whatever,*[52] Houellebecq writes on page 91: "The attractile drives are unleashed around the age of thirteen, after which they gradually diminish, or rather they are resolved in models of behavior which are, after all, only constrained forces. The violence of the initial explosion means that the outcome of the conflict may remain uncertain for years; this is what is called a transitory regime in electrodynamics."

What literary effect is Houllebecq going for by using the word "electrodynamics," a barbarous word for any nonphysicist reader?

And on page 125: "Of all economic and social systems, capitalism is unquestionably the most natural. This already suffices to show that it is bound to be the worst. Once this conclusion is drawn it only remains to develop a workable and consistent set of concepts, that is, one whose mechanical functioning will permit, proceeding from facts which reinforce the predetermined judgment, the way that bars of graphite can reinforce the structure of a nuclear reactor."

Come on now—the graphite he refers to here is actually used for slowing down the neutrons, not for supporting the structure! Houellebecq isn't a man of learning; he plays one, and he's doing well at it, judging by the success of his books. Like others, he takes advantage of the generally accepted idea that the laws of the Universe are mysterious, since if they weren't, there would be little difference between Man and God. Then, to seem as though he possesses an unusual facility to grasp these laws, he writes in a fashion that is hard to read or, better yet, completely incomprehensible. And, besides being fraudulent, this kind of behavior promotes a disastrous image of science!

That reminds me, in a different vein, of the thousands of students at Harvard who piled one on top of each other to hear Stephen

Hawking speak in the fall of 1999, due to a collective infatuation I happened to share with them. Each year the Physics Department at Harvard invites five prestigious speakers to give a series of talks called the Loeb Lectures. I had the good fortune to be one of these lecturers, but the bad luck to have to speak right after Hawking. Hawking holds Newton's prestigious chair at Cambridge. He is a physicist who became very famous for his theory of black holes, and especially for a series of bestselling books, including A *Brief History of Time*. In addition, since the age of twenty he has suffered from an incurable degenerative motor neuron disease, which has progressively bound him to a wheelchair, in which he lives with difficulty, since it's very rare for his muscles to respond to his will. He is able to move a single finger, with which he commands a voice synthesizer; nonetheless, though he manages to smile and maintains a look of wonderful vitality, holding a conversation with him in person leaves one with the unforgettable impression that one is speaking directly with Hawking's brain.

Faced with his misfortune, Hawking fights to stay alive and productive. He has hired two nurses to help him in his daily life, and he is constantly surrounded by several students who help him to work. His arrival at Harvard was an event, to say the least. For thousands of students from all disciplines, this uncommon man, necessarily an oddity, had come to explain the mysteries of the universe. The largest lecture hall at the university, with around 3,000 seats, was full to bursting. They were supposed to open another room as well, so that people could watch a video transmission of the lecture. At the appointed hour, an electric wheelchair came to the stage, tracked by spotlights. And Hawking began. He wanted to explain to us why the Universe was both infinite and bounded. So he launched from the very beginning into unusually complex developments in the topology of "d-branes" in "de Sitter" spaces. I later learned that "d-branes" were a generalization of membranes in d-dimensional spaces, where d is greater than three or four, the spatial dimension with which we are more or less familiar. I was especially surprised to see the crowd become more and more fascinated as the talk became more and more incomprehensible.

The fervor didn't die down for the next session. I should add that, since he's a good comedian, Hawking embellished his talks with funny drawings. Every fifteen minutes or so, each time the difficulty of what he was saying required a rhetorical transition, he would pause, and show us a cartoon of himself meeting God and asking him something like: "So, that's where I'm at, but what would you think if I tried this?"

Since English doesn't distinguish between the second person formal and informal, I don't know how familiar Hawking felt he really was with God, but he played cleverly with his character. He had indeed met God, he seemed to be saying, and he could understand the mysterious origins and the structure of the Universe only because he had chatted with God. Is that what had allowed him to survive his terrible physical ordeals? In the end, the crowd wanted more, and Harvard organized an additional lecture, this time charging admission, and in the largest space in the city: 5,000 seats if memory serves. It was expensive, and full.

Two weeks later, it was my turn. How could I put myself in anything like the same league by announcing a lecture on "Crystallization and Nucleation?" Fortunately, the department reserved only the physics amphitheater, and there were about a hundred of us present. As a result of my stage fright, I adopted the only attitude that seemed possible: I was going to be simple and pedagogical; each of my words, each of my claims, each phenomenon I described would be understood by everyone and in full detail; and for each question I solved, I would bring up a related one which was unsolved. I didn't attract any non-physicist students, but the small group was attentive. Some young researchers had come from nearby MIT, and they still tell me about how engaging my talk was.

Hawking was at the gala dinner that followed my first lecture. During the coffee afterwards, we ran into each other, and he told me—at least, had his computer tell me, while he smiled—that he was sorry to have been unable to attend my lecture. Of course I responded that it was no problem, and that I well understood that he must have been busy elsewhere. Hawking is a great scientist, and I perfectly understand that the costs required for his survival are such that he must earn money as a public figure to meet his needs. (Not at all like Houellebecq—especially since Hawking understands all of the science he discusses.) Nonetheless, I wish all scientists would try to be understood by the greatest possible number of people when they leave their specialized niches and communities. That way, science would be what it should: a source of reflection, not an instrument of power.

If you think that science is incomprehensible, especially contemporary science—and especially physics; if you're afraid of looking ridiculous if you don't understand; if you still feel the sting from a dogmatic teacher and the way he graded you; if you think that things are either true or false—and are so now and forever; if you don't dare to ask naïve questions; if you are a victim of the intellectual terrorism called "scientism," I've written this book for you and have tried to bring you to the frontiers of science, with a few anecdotes along the way, in the hopes that you might understand why scientific research is my passion—and perhaps even share it.

Notes

1. A rough equivalent might be an engineer for the Bureau of Land Management. [tr.]

2. I highly recommend Dava Sobel's admirable biography, written on the basis of letters to Galileo from his eldest daughter—a masterpiece: *Galileo's Daughter* (Walker & Co., 1999).

3. Sirius, in the Canis major constellation, is about 10 light years away from the Earth, that is, a hundred trillion kilometers; but, on average, there is less than one star every 10 lightyears. That's because there is empty space between galaxies. In addition, the density of stars is not constant, but it would be too involved for me to get into a description of the exact distribution of stars in the universe here, which wouldn't change my reasoning at all. Specialists, rest assured; I am not ignoring the fact that the Universe is an aggregate of stars that could be called "fractal."

4. Our galaxy is the Milky Way, the great heap of stars that crosses the sky from the north-east to the south-west, and which we see this way because the Earth isn't at its center: our galaxy is made up of stars which rotate in a disc-shape, and we see it from the inside of its plane. If you go to the southern hemisphere you will even see the center of our galaxy, in the middle of the Milky Way—a dark-looking region, since the cloud of dust there absorbs the light from the central stars.

5. He clarified a bit later that they also needed to be transformed into their mirror images: that is, particles spinning to the left needed to be replaced by antiparticles spinning to the right. Chirality again.

6. Ludwig Boltzmann, an Austrian physicist and philosopher born in Vienna in 1844, founded statistical physics by coming up with the kinetic theory of gases. Severely criticized from 1880 on, and ill, he committed suicide near Trieste in 1906.

7. The number following the name of an isotope is the sum of the number Z of protons and the number N of neutrons which the nucleus contains. The name of an element corresponds to the number of protons, but each element exists in the form of different isotopes which each correspond

to a different number of neutrons. Basically, if Z is about equal to N, as in ordinary carbon—carbon 12 ($Z = N = 6$) or carbon 13 ($N = 7$)—the isotope is stable. But if N is too large or too small, the isotope is unstable and radioactive. Thus carbon 14 disintegrates into nitrogen 14 by emitting an electron and another particle called a "neutrino."

8. It would be practically impossible to repair all the misinformation which accompanied that cloud's passage over France. The "Central Service for the Protection against Ionizing Radiation", an official service directed at the time by Professor Pellerin, did note a "slight increase in radioactivity" in France, which he communicated to the public and confirmed on May 2. In Switzerland, a number of women decided to have abortions in June 1986, following alarmist messages which worried them. In France, on the contrary, no such panic took place. At the end of an interview with the French managers of the Europa Bridge, between Strasbourg and Kehl, a journalist concluded: "It's as if the cloud didn't cross the Rhine." The myth was born. Some malicious people put that in the mouth of Professor Pellerin as "the Chernobyl cloud did not cross the Rhine," to make him look ridiculous. It's generally impossible to repair slander.

9. To be named Kurie and work on radioactivity . . . That reminds me that I have an American colleague named Einstein, an excellent theorist working on crystalline surfaces and other questions in statistical physics, though his first name isn't Albert, all the same. I never dared to ask him whether he sometimes felt condemned to excellence.

10. Many researchers in this very active domain are working carefully on the definition of units in physics, such as that of the new meter, which is much more precise than the one made from platinum.

11. The École Polytechnique is run by the military, and its students would therefore normally participate in military parades, e.g., on Bastille Day, the 14th of July.

12. I hope specialists in musical acoustics will excuse me from entering into affairs of timbre and harmonic distribution. That isn't what interests me here, and we'll get lost if I talk about everything at once. I hope they will also excuse me from entering into questions, however interesting, related to the ways of tuning instruments, i.e., temperament.

13. A nanometer (1 nm) is equal to a billionth of a meter (10^{-9} m). A thousand nanometers is approximately the thickness of a human hair.

14. Richard Feynman, a Nobel Prize winner in physics in 1965 for his work on quantum electrodynamics, once admitted that "no one really understands quantum physics."

15. It isn't the case, for example, with butterfly wings or with soap bubbles, where it is caused by interferences.

16. Wolfgang Pauli was born in Vienna in 1900 and worked with Max

Born, then with Niels Bohr, before being named professor at Zurich; he then spent several years at Princeton and, finally, returned to Zurich after the war.

17. A. Sokal and J. Bricmont. *Impostures intellectuelles*. Paris, Odile Jacob, 1997 [English edition—Translated as *Fashionable Nonsense* (Picador, 1999)].

18. S. Balibar, "The discovery of superfluidity," *Journal of Low Temperature Physics* 146: 441 (2007).

19. A poise is a unit of viscosity which gets its name from Jean-Louis Marie Poiseuille, a French doctor and physicist who studied the flow of blood at the beginning of the nineteenth century; the viscosity of water is on the order of one-hundredth of a poise.

20. "Die Theorie ist hübsch, aber ob auch etwas wahres dran ist?" Letter to Paul Ehrenfest, November 29, 1924.

21. L. D. Landau, "The Theory of the superfluidity of helium II." *Journal of Physics (USSR)* 5: 71 (1941).

22. On this subject, see our exchange of letters in the *Bulletin de la Société française de physique*, 129: 18, 2001, which also shows that scientific controversies need not ruin friendships.

23. Lev Pitaevskii, "Fifty years of Landau's theory of superfluidity," *Journal of Low Temperature Physics*, 87: 127 (1992).

24. This prestigious prize was later won by John Bardeen, as well as Abrikosov, Leggett, and several other Nobel winners, but also by less famous physicists—including myself as I was finishing this book. John Bardeen got one Nobel Prize for the invention of the transistor, then a second one for the theory of superconductivity. On that occasion, he considered his debt to London to be so great that he donated part of his prize money to a foundation which today allows the winners of the London Prize to receive a check. Having won this prize myself, I am obviously no longer able to be astonished at the choices of its jury. In an email from June 2005, Laszlo Tisza informed me that he himself had suggested Landau for the London Prize. Clearly, the great man didn't know how to hold a grudge.

25. Pierre Curie laid out his principle of symmetry thus: "When certain causes produce certain effects, the symmetrical elements of the causes must be reproduced in the effects. [. . .] When certain effects reveal a certain asymmetry, this asymmetry must also be found among the causes which gave rise to it." *Journal de Physique*, 3rd ser., 2 (1894).

26. The word "crystal" comes from the Greek *krustallos*, which means a piece of ice ("cryogenics," the technology of cold temperatures, shares the same root). By extension, it designates "rock crystals," that is, quartz and semiprecious stones.

27. I also remember having shocked several flautists by writing in an

article on the acoustics of flutes (*La Recherche* 188: 36 [1981] that the sound was related to the shape of the pipe, not its material. In his book, the great Jean-Pierre Rampal wrote that you could hear the wood of baroque flutes and that the sound of a golden flute was obviously better than the sound of a silver flute. But the shape of baroque flutes is completely different from that of modern metal flutes. Moreover, the flautist has two hands on the instrument, the air inside of which is all that vibrates, not the solid body.

28. S. Balibar, H. Alles, and A. Ya. Parshin, The surface of helium crystals, *Review of Modern Physics*, 77: 317 (2005).

29. English translation of *Il Sistema Periodico*, copyright 1984 by Schocken Books, Inc.

30. Introductory text for the reform of high school physics programs, 1999.

31. S. Balibar, J. Buechner, B. Castaing, C. Laroche, A. Libchaber, "Experiments on superfluid ^{4}He evaporation," *Physical Review B.*, 18: 3096 (1978).

32. Another Nobel prize winner, this one in 1977.

33. Daniel Bernoulli and his brother Nicholas were born in Bâle; like their friend Leonhard Euler, they learned mathematics from their father, Jean Bernoulli, a friend of Leibniz and the pupil of his own father Jacques Bernoulli. Leonhard Euler's father, Paul Euler, had also studied mathematics with Jacques Bernoulli. What families!

34. The daughter of a Lithuanian serf and a prisoner of the Russian army at age 18, she became Peter the Great's mistress, whom he married in 1712. Upon his death (1725), she reigned briefly over Russia and founded the Academy of Saint Petersburg before dying as well. Not to be confused with Catherine II (the Great), who reigned from 1762 to 1796 and enjoyed spending time with Voltaire and Diderot.

35. Born in Budapest in 1881, Theodor von Karman worked at Aachen before emigrating to California and, from 1930 to 1960, playing an important role there in the development of fluid mechanics, especially its applications to aeronautics.

36. Gérard Jorland, *Les paradoxes du capital* (Paris: Odile Jacob, 1995).

37. *The Holy Bible*, King James Version.

38. S. Douady and Y. Couder, *Physical Review Letters*, 68: 2098 (1992).

39. That's what the group of French mathematicians who did the same, under the name Nicolas Bourbaki, suggests.

40. "Mathematical recreations," *Scientific American*, January 1995, 76.

41. Türing would become famous most of all for inventing the "Türing machine," a concept of a computer.

42. See for example the 2007 report of the Intergovernmental Panel on Climate Change on http://www.ipcc.ch/

43. The law passed on August 4, 1994.

44. The law, of course, does not extend the reciprocal right to foreign authors, Germans for example, to insert summaries in German.

45. I'm well aware that in other disciplines, globalization takes different forms or the evolution doesn't come as quickly. Specialists in French history, for example, have less need to express themselves to audiences which don't understand French. The case of mathematics is also often cited, the expression of which in French better withstands anglicization. Perhaps, though, that is simply because the equations are written the same way in every language (more or less).

46. For readers who weren't around during that era, a euro in 2005 was worth about 6 francs in 1995. I don't know if these fines are indexed to the cost of living.

47. Claude Allègre, "La clé de l'anglais," *Le Point*, May 21, 1994.

48. "L'anglais pour défendre la langue française," *La Recherche*, September 1992; "Les chercheurs français face à la loi Toubon," *La Recherche*, November 1994.

49. R. Balian, "Le physicien français et ses langages de communication," presented at the conference "La langue française et la science," Consulate General of France, Quebec, March 19, 1996.

50. Renée Balibar, *Le Colinguisme*, "Que sais-je?" series (Paris: PUF, 1993); *Histoire de la littérature française*, "Que sais-je?" series (Paris: PUF, 1993).

51. Originally published as *Les Particules Elementaires*, Flammarion 1999. trans. Frank Wynne (New York: Vintage, 2001) in the UK by William Heinemann), p. 356.

52. Originally published as *Extension du domaine de la lutte*, (Maurice Nadeau, 1994); *Whatever*, trans. Paul Hammond (London: Serpent's Tail, 1998).

Index

Abrikosov, Alexei, 85
absolute temperature, 49
absolute zero, 49
Adam, 133
age of the Earth, 32
age of the Universe, 7
Allègre, Claude, 168
Allen, Jack F., 75
alpha particles, 32
Anderson, Carl, 21
anti-hydrogen, 22
antimatter, 20
anti-particles, 22
apples, 55, 133
archaeology, 31
Archimedes, 122
art, 33
atoms, 55
Auschwitz, 160

BaBar, 25
background radiation, 12
baker, 125
Balian, Roger, 169
Bardeen, John, 66
Bauer, Edmond, 82
BCS, 66
Becquerel, Henri, 28
becquerels, 26
Berners-Lee, Tim, 74
Bernoulli, Daniel, 115
Bhopal, 152
Bible, 132
bicycles, 72

Big Bang, 6, 8, 12
Big Crunch, 9
billiard balls, 24
billiards, 124
Binnig, Gerd, 55, 101
biologists, 109
biology, 195
biomass, 158
black holes, 12
Bohr, Niels, 44
boiling, 114
Boltzmann, Ludwig, 24
books, 170
Bose, Satiendra Nath, 40
Bose-Einstein condensation, 42
bosons, 47
Bourbaki, Nicolas, 137
Bravais, Auguste, 138
Bravais, Louis, 138
Brézin, Edouard, 71
Bricmont, Jean, 74, 174
Broglie, Louis de, 49
butterflies, 58, 127
butterfly effect, 124

Cambridge, 75, 76
Canup, Robin, 35
carbon 14, 26, 30
carbon 60, 52, 65
cards, 24
Catherine I of Russia, 115
Caupin, Frédéric, 116
cavitation, 115
CERN, 74, 90

cesium, 72
cesium 137, 27
CH_4, 150
chaos, 120, 130
charm, 106
chemistry, 59
Chernobyl, 27, 151
Chicago, 146
chirality, 18
chromosomes, 143
Ciccone, Vincent, 71
climate, 203
clocks, 57, 123
clouds, 28, 58
CO_2, 150
Cockroft, John, 79
coffee, 72
Cohen, Daniel, 128
coherence, 62
Cole, Milton, 72
color, 57
common ancestor, 19
compass, 131
complexity, 68
conferences, 166
Constitutional Council, 165
convection, 122
Cooper, Leon, 66
Copernicus, 5
cosmic rays, 21, 30
cosmological constant, 15
cosmonauts, 73
Couder, Yves, 134
CRIIRAD, 27
curie, 27
Curie, Marie, vii, 28, 30, 73
curiosity, 133
curvature of space, 11

dams, 157
dark energy, 13
dark matter, 10
dark night, 6
dating, 34
degrees Celsius, 49

degrees Kelvin, 49
density, 54
deterministic laws, 125
deuterium, 154
diagonals, 72
dinosaurs, 35
Dirac, Paul, 20
DNA, 18, 29
Doppler, Johann, 8
Douady, Stéphane, 134
Dutch, 162
dynamic systems, 143

$E = mc^2$, 20
Earth's magnetic field, 130
Earth's mantle, 33
earthquakes, 122
eclipses, 37
ecliptic, 38
École normale supérieure, 71
economics, 129
Ehrenfest, Paul, 82
Einstein, Albert, 11, 15, 30, 40, 66, 173
El Niño, 127
electric field, 60
electrical resistance, 66
electron, 47, 55
email, 73
embryo, 144
encoding, 68
energy, 148, 149
energy conservation, 156
English, 160
ephemerides, 5
Euclid, 137
Euler, Leonhard, 52, 115
Eve, 133
exoplanets, 37
expansion of the Universe, 10, 14
extraterrestrials, 37

facets, 98
Feigenbaum, Mitchell, 122
Fermi, Enrico, 146

fermions, 47
Feynman, Richard, 21, 57
Fibonacci, 135
Fizeau, Armand, 8
flute, 46, 56, 96
fountain effect, 81
French, 169, 172
frustration, 97
fuel cells, 159

galaxies, 10
Galileo, 3, 4
Gamow, George, 12
general relativity, 11
Gennes, Pierre Gilles de, 72, 101
Germany, 160
God, 133, 175
Golden number, 141
GPS, 12
gravitation, 9
greenhouse effect, 150
Gulf Stream, 126
Guthmann, Claude, 100

Hadamard, Jacqueline, 83
Haigneré, Claudie, 156
half-life, 31
hands, 17
Harvard, 175
Hawking, Stephen, 175
Heisenberg, Werner, 55, 83
Heitler, Walter, 82
helium, 48, 72, 102
helium crystals, 100
Hiroshima, 28
Homo sapiens, 32
honey, 119
Houellebecq, Michel, 174
Hubble, Edwin, 7
Hugo, Victor, 33
hydrogen, 55

images, 42
information, 70
infrared, 150

Institut Curie, 71
Institut Henri Poincaré, 82
integrated lasers, 47
interactions, 23
interferences, 62, 67
International Space Station, 156
Internet, 83
invisible matter, 10
iridium, 36
Iron Curtain, 169
ITER, 154

Jacques, Jean, 33
Jones, Harry, 81
Jorland, Gérard, 129
Jupiter, 3, 4

Kamerlingh Onnes, Heike, 66
Kapitza Institute, 72
Kapitza, Piotr, 72, 75
Karman, Theodor von, 118
Keeson, Willem, 75, 100
Kelvin, Lord, 43
Kepler, Johannes, 137
Keshishev, Konstantin, 101
Ketterle, Wolfgang, 50
Kurie, F. N. F., 30
Kyoto treaty, 159

lambda point, 75
laminar, 119
lamp, 62
Landau, Judd, 101
Landau, Lev, 75, 78, 81
Langevin, Paul, 82
languages, 170
laser, 62
Laskar, Jacques, 123
Latin Quarter, 71
Le Corbusier, 137
leaks, 73
Lee, T. D, and Yang, C. N., 23
leeks, 20
Leggett, Anthony J., 87
Leonardo da Vinci, 137

leukemia, 28
Levi, Primo, 103
Lhomond, rue, 73
Libby, Willard Franck, 30
Libchaber, Albert, 72, 121
Lifshitz, Evgeny, 86
light, 58
Lipson, Steve, 101
liquid helium, 65, 75
London prize, 89
London, Fritz, 81
Lorenz, Edward N., 124
LUCA, 19
Lucy, 32

magnetic field, 60
man, 32
marching, 50
Mars, 5
Masson, Jean-Louis, 160
Maurer, Jean, 121
Maury, Jean-Pierre, 2
medical imaging, 48
memory, 124
Mendeleev, 77
Mercury, 11
mercury, 66
meteorology, 126
methane, 158
Middle East, 151
mirror, 23
Misener, A. D., 75
MIT, 83
molecules, 59
Moon, the, 3, 34
MOX, 152
musical acoustics, 120

Nature, 75
Navier Stokes Law, 125
negative pressure, 13, 115
neutrinos, 10
neutrons, 155
Newton, Isaac, 55
noble gases, 102

noise, 121, 129
non-linearity, 129
Nozières, Philippe, 103
nuclear reactions, 26
nuclear waste, 29
numbers, 124

Ohm's Law, 43
oil, 149, 159
Olbers, Wilhelm, 6
Oppenheimer, Robert, 21
oranges, 102
orders of magnitude, 147
original sin, 133
Oshima, Aiko, 71
ozone, 30

Parshin, Alexandr, 101
Pasteur, 19
Pauli, Wolfgang, 59
Pauli's exclusion principle, 59
Peierls, Rudolf, 76
peloton, 117
pencil, 23
Penzias, Arno, 13
Perrin, Jean, 82
photons, 22, 48
photosynthesis, 158
phyllotaxis, 138
Physical Review, 163
piano tuners, 146
pianos, 146
Pieranski, Pawel, 103
pinecones, 134
Pitaevskii, Lev, 87
planet, 3, 10, 55, 58
plutonium 239, 152
Poe, Edgar Allan, 12
Poiseuille, Jean-Louis Marie, 76
Pomeau, Yves, 87
positrons, 21
pots, 33
predictability, 124
probability, 62
protons, 55

publications, 164
puff pastry, 125
Pythagoras, 137

quantum boxes, 47
quantum computers, 67
quantum excitement, 56
quantum mechanics, 54
quantum physics, 44
quantum state, 44, 57

radioactive waste, 152
radioactivity, 26
Raffarin, Jean-Pierre, 154
referees, 78
relativity, 20
renormalization, 112
reproducibility, 79
Rohrer, Heinrich, 55, 101
Rolley, Etienne, 100
Rorschach test, 16
Rouan, François, 98
rough, 102
rubidium, 33
Ruelle, David, 123
Rutherford, Ernest, 32, 77

Saam, William F., 72
Saint-Geneviève, montagne, 71
Sakharov, Andrei, 24
satellites, 126
Saturn, 6, 35
scanner, 48
Schrieffer, Robert, 66
Schrödinger, Erwin, 60
Schrödinger's cat, 60
Schrödinger's equation, 20
Schumann, Robert, 161
Science, 42
science fiction, 38
scientism, 173
sensitivity to initial conditions, 124
Sinai, Yakov, 124
sky, 9
slander, 27

soap bubbles, 58
sodium, 40
Sokal, Alan, 74, 174
solar cells, 157
solar energy, 157
spin, 21, 48, 60
spinning tops, 60
spirals, 134
Stalin, 72, 77
standard model, 25
stars, 5
Stewart, Ian, 140
Sun, the, 30, 34, 55, 154
sunflowers, 134
superconducters, 48
superconductivity, 66
superfluidity, 66, 75
Superphénix, 153
symmetry, 17
symmetry-breaking, 23

tables, 54
Takens, Floris, 123
Tanner, Leonard, 101
teaching, 172
telescope, 2
television, 163
Teller, Edward, 83
temperature, 49
thermonuclear fusion, 154
thorium, 154
Three Mile Island, 151
Tisza, Laszlo, 81
Toubon, Jacques, 164
Toubon's law, 164
trajectory, 55
transistor, 68
translation, 168
traveling salesman, 68
trees, 115
Treiner, Jacques, 72, 107
tritium, 154
turbulence, 119
Türing, Alan, 143
two fluids model, 84

uncertainty principle, 55
Universe, 6, 9, 177
uranium, 33

vacuum, 73
Venus, 3
violin, 44
viscosity, 76

water, 37
waves, 61

Weyl, Hermann, 21
"What Do I Know?", 172
whirlpools, 115
wild ducks, 119
Wilson, Robert, 13
windmills, 148
wood, 149
words, 74
Wu, C. S., 24

Zeilinger, Anton, 65

"I am among those who think that science is a thing of great beauty. A scientist in his laboratory is not just a technician: he is also a child confronted with natural phenomena which dazzle him like a fairy tale.

"We should not let others believe that all scientific progress can be reduced to mechanisms, machines, and gears, though they have their own beauty as well . . .

"I do not think either that in our world, the spirit of adventure is likely to disappear. If I see anything vital around me, it is precisely that spirit of adventure, which seems impossible to uproot, and resembles curiosity . . ."

—Marie Curie